Peter Wohlleben spent over tw
forestry commission in Germany i
of ecology into practice. He nov
friendly woodland, where he wor.
forests. He is the author of numerous books about trees,
including the bestselling *The Hidden Life of Trees*.

To learn more about Peter and his books, visit his website
at peterwohllebenbooks.com.

Jane Billinghurst is a nature lover, master gardener, editor,
translator and author of six books. She has translated and
edited several books by Peter Wohlleben, including the *New
York Times* bestseller *The Hidden Life of Trees*. She lives in
Anacortes, Washington, next to 2,800 acres of community
forest lands.

Praise for *How Trees Can Save the World*

'Wise and thought-provoking' *Observer*

'If we don't learn to leave the trees alone, the trees will
eventually be alone anyway – but without us. Wohlleben
brilliantly and readably shows us how urgent and how
hard it is to do nothing.' *Guardian*

'Wohlleben . . . offers a pointed critique of harmful
forestry practices and urges humans to let trees heal
themselves . . . [T]he insights into trees' surprising abilities
captivate . . . Nature lovers should take note' *Publishers Weekly*

'Another love letter from Wohlleben to the green world . . .
[*How Trees Can Save the World*] makes the case for how we should
allow forests throughout the world to regrow and in the process help
heal not only the climate but us, as well.' Lydia Millet, *Oprah Daily*

'Bestselling environmental writer Wohlleben draws largely
on observations of his home territory in Germany in this
thoughtful look at how trees learn and adapt . . . the forest is
often the last recipient of humanity's benevolence. The author
is in his element here as a gentle purveyor of knowledge that
provides a new perspective on a crucially important topic.
His many fans will be enthused, and new readers will appreciate
entering Wohlleben's evocative world.' *Booklist*

How Trees Can Save the World

Peter Wohlleben

Translated by Jane Billinghurst

DAVID SUZUKI INSTITUTE

WILLIAM
COLLINS

William Collins
An imprint of HarperCollins*Publishers*
1 London Bridge Street
London SE1 9GF

WilliamCollinsBooks.com

HarperCollins*Publishers*
Macken House
39/40 Mayor Street Upper
Dublin 1
D01 C9W8
Ireland

First published in English by Greystone Books in 2023

Originally published in German as *Der lange Atem der Bäume: Wie Bäume lernen,
mit dem Klimawandel umzugehen—und warum der Wald uns retten wird,
wenn wir es zulassen,* copyright © 2021 by Ludwig Verlag, Munich, part
of the Penguin Random House Publishing Group GmbH
This William Collins paperback edition published in 2024

1

Peter Wohlleben asserts the moral right to be identified as the author of
this work in accordance with the Copyright, Designs and Patents Act 1988

A catalogue record for this book is available from the British Library

ISBN 978-0-00-844724-3

Editing by Jane Billinghurst
Copyediting by Brian Lynch
Proofreading by Alison Strobel
Indexing by Stephen Ullstrom

Printed and Bound in the UK using 100% Renewable Electricity at CPI Group (UK) Ltd

This book contains FSC™ certified paper and other controlled
sources to ensure responsible forest management.

For more information visit: www.harpercollins.co.uk/green

Contents

PART II
When Forestry Fails

PART III
Forests of the Future

Introduction

THE FUTURE OF FORESTS and the future of humanity are inextricably entwined. I'm not saying this for dramatic effect; it is simply a fact. As bleak and frightening as these words sound, they actually offer a great deal of hope. Trees are so adept at creating communities that many of them can cope with the current level of climate change. They are also our best option for removing greenhouse gases from the atmosphere, doing so much more efficiently than any technological fix we could ever come up with. They also markedly cool climates locally and significantly increase rainfall.

Trees, by the way, are doing all this not for us, but for themselves. Like people, trees don't like conditions to be too hot or too dry. Unlike people, they can turn the thermostat back down a bit. That said, beeches, oaks, and spruce are not born with all the skills they need to effect these changes. On their long journey to becoming old trees, they must learn how to adapt. Not every tree succeeds, because these enormous plants are just like people: there are huge variations between individuals, and they don't all learn at the same speed or draw the right conclusions from their life experiences.

In this virtual journey through the forest, I'll show you how you can watch trees learn, why it's not necessarily a problem for beeches or oaks if they drop their leaves in summer, and how you can tell when trees have opted for the wrong survival strategies.

Even though science has made great progress in bringing the hidden life of trees out into the open, the curtain has still barely been lifted. The role played by tiny organisms such as bacteria and fungi, for example, remains largely unexamined, primarily because most species are as yet undiscovered. And yet these tiny life-forms are as important to trees as our gut microbiome is to us: without these microorganisms life would not be possible for either trees or humans. Fascinating new details from this hidden world reveal that each tree is an ecosystem in its own right, like a planet populated by an infinite number of amazing life-forms.

When we step back, the big picture also reveals surprises. Forests create atmospheric rivers carrying water in cloudbanks that travel thousands of miles into the interior of continents, dropping rain on regions that would otherwise be deserts.

Trees, therefore, are not life-forms that stand there and suffer as human activity changes the global climate. Rather, they are creatures rooted in their environments that react when conditions threaten to get out of control.

There are, however, two things trees need to be able to adapt successfully: time and being left alone. Every intervention sets the forest ecosystem back and prevents it from establishing a new equilibrium. If you have walked by clearcuts in Germany recently, which are some of the largest in decades, you will probably already have noticed the extent to which modern forest practices interfere with the trees' ability to recalibrate. But there is hope! Forests return quickly and vigorously when they are allowed to grow back on their own. All we need to do is to accept that we cannot create forests; the best we can do is to set out plantations. A better way for us to help trees is to step aside and allow natural reforestation to take its course. If we maintain an appropriate level of humility about our abilities and optimism about the power of nature to heal itself, the future can be one thing above all—green!

PART I

The Wisdom of Trees

1

When Trees
Make Mistakes

IN HOT, DRY SUMMERS, trees have big problems. They cannot escape to the shade, and they cannot take a sip of water to cool themselves down. Indeed, they cannot react quickly in any way. And because they're so slow, it's all the more important for them to choose the right coping strategy. But what is the right strategy and what happens when a tree makes a mistake?

I had a ringside seat to observe this from the academy I established in Wershofen in the German state of Rhineland-Palatinate to educate people about forests and help support struggling forests around the world. One side of North Street is lined with horse chestnuts. In the dry summer of 2020, the horse chestnuts behaved much like many trees in Europe that year: their leaves began to take on the colors of fall in August, which is far sooner than normal.

Horse chestnuts have been having a particularly challenging time in many parts of Europe for years. Shortly before 2000, horse-chestnut leaf miners advancing northward reached the trees in Wershofen. These small, light-brown moths are native to Greece and Macedonia, the horse chestnuts' original homeland. Up until the moths' arrival, like many other imported plants, the horse chestnuts in

Wershofen had been leading charmed lives. Although the ecosystems in countries like Germany are not perfect for these trees—it really is a bit too cold for them this far north—nevertheless, our chestnuts settled in nicely. The parasites that plagued them back home had not yet made it to the trees' new location and being slightly colder in winter was a small price to pay for a life without leaf miners. Then, about forty years ago, things started to change. That was when the moths began to follow their prey north, and eventually they, too, arrived in Wershofen.

Leaf miners do just as their name suggests: the caterpillars "mine" tunnels in leaves. The female moths lay their eggs on the surface of the leaves, and after they hatch, the caterpillars eat their way inside. Small brown, wavy lines show where the moth babies have been happily chowing down—happily, because living inside a leaf offers good protection from hungry birds. The mined sections of the leaves dry out, and the caterpillars continue to eat. As summer progresses, the foliage looks increasingly ragged, especially as the first round of egg laying is often followed by a second.

The leaves on the trees along North Street were therefore already damaged when, after several hot days, the drought settled in. In a situation like this, chestnuts react just as all trees do: they shut down photosynthesis and wait. The trees have even less of an idea than we do of how long a dry period like this might last and therefore it makes sense for them not to panic right away.

The trees' first response is to close the millions of tiny mouths, the stomata, located on the undersides of their leaves. Trees use these tiny mouths to breathe, just like we do, and, just like us, trees expel water vapor with every breath.

The water vapor cools their surroundings as it evaporates, and the green giants actively manage this process to make hot summer days more bearable.

When the roots signal all the moisture in the soil has been used up, the trees close these countless mouths. However, when the stomata are closed, photosynthesis stops because carbon dioxide is no longer entering the leaves. Without water and carbon dioxide, the trees can no longer use sunlight to produce sugar. At this point, the trees begin to consume the sugar reserves they were hoping to increase so they could survive their long winter sleep.

Despite the shutdown, the trees continue to lose a minimal amount of moisture through their leaves, roots, and bark. If the drought continues, they now employ a second strategy: they discard some of their leaves. Like their fellow green giants, the chestnuts lose their leaves from top to bottom, and the first leaves to fall are those way up in the canopy, farthest from the roots. Trees expend a great deal of energy transporting water to their crowns, and because they can no longer stockpile sugar, they need to conserve energy. If dropping the topmost leaves doesn't do the trick, and if the rain still doesn't come, the trees continue to drop leaves gradually from the top down until finally, as early as August, they stand there with completely bare branches.

In 2020, almost none of the beeches, oaks, or chestnuts around Wershofen had resorted to this final step. Only a few individuals had given up. Perhaps these trees were particularly anxious and wanted to play it safe. Or perhaps they were growing in especially dry spots. Whatever the reason, by August, the branches on these trees had no leaves left on them at all.

The chestnuts in particular really couldn't afford to lose their leaves, as by summer they had already been weakened by leaf miners. Their leaves, covered in brown patches where the caterpillars had fed, were producing limited quantities of sugar, which meant the trees were starving. Next on the chestnuts' list of challenges was the elevation at which they were growing. North Street lies about 2,000 feet (600 meters) above sea level. The elevation and the harsh conditions in the Eifel, a low mountain range in central Germany, make for a short growing season and the window to produce sugar is tight. Trees need to produce enough not only to keep going day to day, but also to see them through their winter rest period and get them kick-started the following spring. In the Eifel, far from their original home, this was already a struggle for the chestnuts. And now they were dealing with the third dry summer in a row, during which every last drop of water in the soil had clearly been used up.

Under normal circumstances, when trees encounter conditions like these, they simply advance their seasonal shutdown and drop their leaves in September—which is what the beeches in my forest typically do. Although the beeches look dead, they leaf out again in the spring and try to make up for time lost the previous year. Most of the chestnuts do this too. The anxious ones that dropped their leaves in August had definitely jumped the gun and employed this survival strategy too soon.

On August 31, 2020, the weather gods had a change of heart. In a small area on the northern edge of the Eifel, clouds darkened the sky. It rained for hours, dumping about a quarter of an inch of rain (60 liters of water per square meter). This wasn't nearly enough to replenish water reserves in the drought-stricken soil, but the upper few inches got a bit damp, and I hoped it would be enough to give the trees some respite.

The panicky chestnuts' reaction over the next few days took me by surprise and initially made no sense: they began to blossom. A tree that is already short on sugar shouldn't be putting energy into reproduction, especially since producing blossoms isn't going to pay off this late in the year. Even if the flowers were fertilized, there would be no time to develop seeds and fruit before the onset of winter.

I was on my way back to the forest academy with a group of forest guides in training when they drew my attention to the chestnuts' behavior. We stopped to take a closer look, and it immediately became clear what was going on. Not only were the trees blooming, but they were also unfurling fresh new leaves. This solved the mystery. The chestnuts were starving. With fresh green growth on their branches, they could now stock up on sugar and fill their storage spaces. Clearly, the trees couldn't tell if they were growing only leaf buds or all the buds on their branches, including those that produce blossoms. And this was what we were witnessing. I made a short video with my cell phone and uploaded it to my social media page for discussion. I soon learned some chestnuts in other areas were resorting to the same strategy.

An internet search revealed that a few horse chestnuts had unfurled fall blossoms in previous years. I wasn't totally convinced by the explanations that were being offered, which suggested the stress of climate change and attacks by leaf miners and fungi had brought the trees to the brink. The trees, then, were blossoming in the fall in a final, desperate effort to reproduce.[1] On the face of it, the explanations sounded logical. However, they assumed trees are unable to differentiate between seasons. Blossoms in fall won't result in fruit because the few weeks that remain until winter are nowhere near long enough for fruit to ripen. Any tree that pursues such nonsensical behavior squanders energy and adds to its predicament.

Scientists have known for decades that trees calibrate their behavior based on day length and temperature. In other words, they follow the seasons as accurately as we do—and they do so without needing a calendar. And this is where the next head-scratching assertion pops up: chestnuts are mixing up their seasons. The summer drought that interrupted their uptake of water and therefore their ability to photosynthesize had confused the trees so much that when the rain returned in fall, they thought it must be spring.[2] This line of argument makes absolutely no sense because it ignores how evolution works. If horse chestnuts get so easily confused—after all, summer drought is a natural phenomenon that occurs at least every couple of decades—then how have they managed to survive for more than thirty million years? Any life-form that regularly engages in such nonsensical expenditures of energy would be too weak when faced with a crisis and would relinquish its hold on life.

It's not confusion that drives the chestnuts' behavior, it's hunger. But once a tree has started something, it must see it through. It is not simply a case of unfurling new leaves (together with blossoms it has no use for). A series of events has been put into motion that must be followed through to the bitter end—and each step increases the tree's energy deficit. Growing leaves has taken energy the tree cannot afford to lose. It cashed in its last reserves when it spread its solar panels to manufacture sugar. It has also used up the buds intended for spring. The tree has opened its buds too soon, and if it is not to stand completely bare next year, it must now invest in new ones. And even that is not the end of it: as buds and leaves develop on new growth, the chestnuts must also grow new wood to carry them.

Here's the bottom line: a tree that drops its leaves in summer and is then overtaken by hunger in fall must grow not only leaves (and blossoms it doesn't want or need) but also new wood and new buds. This effort is worthwhile only if the new leaves generate more sugar for the tree to store over winter than the tree spent producing all this new growth. Unfortunately, the march of the seasons is not in the frantic trees' favor. In September, the days are considerably shorter, which reduces the time the trees have to photosynthesize. Moreover, areas of low pressure usually move in a few weeks later, bringing with them a great deal of rain. Although this rain recharges the ground, it also obscures the sun. And if that were not enough, temperatures fall and the first overnight frosts make their presence felt.

That October, the other chestnuts along North Street demonstrated how trees should behave. They withdrew energy reserves from their leaves, which turned first yellow and then brown. They didn't waste time because the first cold snap with overnight temperatures dipping below 23 degrees Fahrenheit (minus 5 degrees Celsius) was going to force them to shut down for the winter. After that, orderly leaf drop is no longer possible and more than just the loss of valuable substances in the leaves is at stake. The only way a tree can actively shed a leaf is by forming a layer of corky cells to separate the leaf from the branch. Trees surprised by the winter shutdown end up with brown leaves hanging from their branches. Heavy snowfall gets trapped and weighs down the branches, and large sections of the crowns can snap off—something I've seen happen many times.

In 2020, when most of the horse chestnuts along North Street behaved as they should, a few panicked. While their fellow chestnuts were sporting fall colors, the trees that had panicked stood there valiantly, covered in fresh green foliage because their sugar account was still in arrears. They didn't shed their leaves until after the first hard frosts in the middle of December, which was far too late. Statistically speaking, some trees that behave this way will not survive the winter and will die just before the time comes to unfurl new leaves in spring. This timing might seem odd to you, so let me explain. Shortly before trees green up, water is forced up the trunk to fuel bud break. This is the trees' most impressive feat of strength in the whole year and the moment when the fate of many weakened individuals is sealed. In the case of the panicked chestnuts in Wershofen, I am pleased to report there was a happy ending. Their buds swelled the following spring, and in one final effort they unfurled new leaves and could finally take their time replenishing their supplies of sugar.

Although it's quite common these days to see chestnuts in various locations blooming and unfurling leaves in fall, I haven't noticed this phenomenon in beech forests. Theoretically, a few beeches here and there could make the same miscalculations as the panicked chestnuts. Maybe they don't thanks to better networking when trees grow together in a forest. The beeches supply each other with sugar solution underground via their root network. When needed, trees in a forest assist other trees that are weak or hungry. This might be why stressed individuals don't grow new leaves to photosynthesize—because they can rely on their community to support them. In contrast, chestnuts planted along a lonely country road far from a naturally occurring woodland community are clearly on their own and must struggle to survive without any help from family.

WHILE YOU CAN WATCH deciduous trees reacting to drought, the process is less visible with conifers. This is hardly surprising, as conifers shed their needles unobtrusively in fall, discarding only the oldest generation. Pines, for instance, hold on to three generations of needles: the current year's needles at the branch tip, needles from the previous year right behind them, and three-year-old needles at the back. Spruce retain up to six years' worth of needles, which is long enough even for them—after all that time, the needles are worn out and it's time to let them go. When you see fall colors in conifers, you know something is wrong.

Conifers actively shed needles, just as deciduous trees actively shed leaves, and like deciduous trees, conifers regulate their use of water in times of drought. First, they shut down photosynthesis. Then they shed needles to reduce the area losing moisture. In these last few years of drought, I had a front-row seat to observe the conifers around my yard at the forest lodge. My wife, Miriam, and I watered the beds close to the lodge to make sure the ornamental plants and vegetables we were growing in our garden didn't simply curl up and die. The hollyhocks and herbs, however, were not the only ones that enjoyed the extra water; the trees growing close by also benefited. Most—but not all—of the

140-year-old pines looked hale and hearty even in the August 2020 heat wave. Any that were not growing right at the edge of the watered beds shed a complete generation of needles prematurely. You could see a huge difference between pines with three years' growth on their branches and those with only two. The old trees with only two looked threadbare. And so my yard and the pines growing there became an open-air laboratory where I could observe as the trees learned from their experience.

SO FAR, WE'VE BEEN concentrating on what happens aboveground in times of drought, but important processes play out belowground as well—in the roots. The roots are probably the most important part of a tree. Cells at the root tips work together to function a bit like a plant brain.[3] The tips test the ground as the roots grow in darkness, continuously monitoring at least twenty different parameters, including moisture, of course, but much more than that. The root tips take note of gravity—after all, the tender tissues need to remain in the soil and not grow up and out of the ground. Light sensors also prevent this from happening. You might think light sensors are unnecessary given that it's always dark underground. Wouldn't gravity sensors be enough? But roots that grow across slopes can end up out of the soil by mistake. It's good, therefore, if the roots have a way of registering when it's getting brighter so they can quickly change course and retreat underground. Roots react with similar horror to toxins. When they come across dangerous particles in the soil, they quickly grow around the problem area (quickly for them, anyway). Based on a diverse range of sensory input, the roots also decide how the tree behaves in general—for example, when it blossoms and how many leaves it grows on its branches.[4]

In dry summers, the most important condition roots actively monitor is, of course, moisture. When this is in short supply, the roots immediately send signals up the trunk and along the branches, telling the leaves to shut their small, mouthlike openings. This puts sugar production—and therefore water use—on hold. Swiss researchers have discovered how this works. They were studying young beeches in the laboratory and had simulated drought conditions in a test setup. The scientists observed that it is indeed the roots that regulate the behavior of the leaves. As the ground dries up, the roots reduce their consumption of sugar—and just when it becomes impossible for them to pump more water up to the leaves, they no longer need to do so: when the roots shut down their intake of sugary liquid from the leaves, sugar supplies back up, which causes the leaves to stop making food.

When trees close the openings in their leaves and stop producing sugar, they switch to their stored supply of food to keep themselves alive. To process their food, the trees need to take in oxygen like we do when we breathe. They are now absorbing oxygen and releasing carbon dioxide, and as a result, a drought-stressed summer forest is no longer a source of oxygen. Once the drought is over, something astonishing happens: the leaves take in more carbon dioxide than normal and produce considerably more sugar. You could say the trees are stuffing themselves to make up for lost time. Their voracious appetites allow them to at least partially make up for the shutdown in sugar production during the drought.[5]

But what happens to the roots during the drought? They need to grow constantly if they are to keep moving through the soil. Under normal circumstances, food flows constantly from the leaves to the tender tissues in the ground. But when photosynthesis stops or, even worse, leaves are discarded, the roots go hungry. This is risky for the trees. If the fine feeder roots die, the trees' ability to take up water will be severely restricted even after the rains return.

When roots die, trees can also lose their balance, as I discovered at the end of 2018. I got up one calm, rainy morning to drive over to my forest academy in the neighboring village. I was at the front door, pulling on my rubber boots, when I heard a strange cracking sound. I peeked outside and watched as a mighty 140-year-old pine slowly leaned over and came crashing down onto one of our woodsheds. When I ran over and looked at the mass of roots, I noticed the fine ones were severely damaged. Dry summers, therefore, affect not only the health of the trees but also their stability.

Before the trees' fine roots start to die, however, the green giants tap into all their sugar reserves, some of which can be very old indeed, as a team of researchers from Finland, Germany, and Switzerland discovered. The team was investigating the age of fine feeder roots—the trees' most delicate roots—by analyzing the carbon they contained. If you measure the number of radioactive atoms in the carbon stored in plant tissue, you can tell how old the tissue is. Cosmic radiation transforms a tiny fraction of the carbon atoms in the atmosphere—more precisely, every billionth atom—into a carbon-14 atom. The half-life of a carbon-14 atom is 5,730 years. Carbon-14 is constantly being created in the atmosphere, but not in plant tissue. Photosynthesis integrates carbon-14 into the plant tissue, where it gradually decays. The proportion of carbon-14 in the carbon stored in the tissue, therefore, diminishes over time. The researchers can tell how old the tissue is by analyzing the ratio of normal carbon to carbon-14. Their research has revealed that the fine roots of native trees in forests in Germany are, on average, eleven to thirteen years old.

If that sounds overly complicated to you, don't worry. There's a much easier way to test the age of the roots: you can slice through them. Tree roots, just like tree trunks, form annual growth rings because they increase in diameter as they grow. Counting these rings led to a real surprise: the roots were ten years younger than the carbon-14 dating suggested, in other words, only one to three years old—and growth rings never lie. According to the researchers, the most likely explanation for the difference is reserves of older food stored within the root tissues. These food reserves age in exactly the same way as the plant tissue ages, and if the tree dipped into these reserves to grow the new fine feeder roots, these roots would have a few years' start on the molecular clock.[6]

You've just read that trees store reserves of food. You likely had no idea that these reserves remain in plant tissue for up to ten years before the trees use them. This was news to me too. The researchers suspect that the development of fine roots from food that has been stored for so long could be a survival strategy used when times are hard. If they are to function as they should, fine roots need to grow even in dry years. As trees cannot manufacture sugar in times of drought, trees that are able to utilize very old supplies of food have an advantage.

This means the old pine in our yard did not necessarily fall because its fine roots had dried out—perhaps it simply did not have enough food stashed away in its tissue that it could draw on in times of need and its fine roots had stopped growing. Likely, however, it simply had not learned how to budget properly and had burned through its reserves of sugar without setting aside an emergency supply. After all, so many dry summers, one after the other, is a rare event this part of the Eifel.

The bad news is that a tree cannot learn from an experience that kills it. The good news is that it is possible for trees to learn the correct strategies to employ and they don't necessarily have to do so through the hard school of life. Trees can protect each other from making life-threatening mistakes, especially when it comes to parent trees and their children. To take a closer look, let's stick with the drought of 2020 and the trees in Wershofen, but this time, we'll step into the middle of a beech forest that has been left to grow at least somewhat naturally.

2

A Thousand Years
of Learning

MODERN EDUCATORS WERE NOT the ones to come up with
the idea of lifelong learning. Trees have embraced this
approach for millions of years. Learning is vital to the survival of beings that can live to be thousands of years old.
Short-lived organisms that reproduce often and in large
numbers can make the adaptations they need to survive
through gene mutation. Under optimal conditions, microorganisms such as Escherichia coli double their numbers as often
as every twenty minutes[1]—something trees can only dream
of doing. It depends on the species, but in extreme cases, it
can take hundreds of years before these giant plants are capable of reproduction. And it takes even fast-growing trees like
birches and poplars at least five years before they bloom for
the first time.

In addition, space needs to be made available before the
next generation can take over in the forest—a vacancy must
be created when a mother tree dies and opens a gap in the
canopy through which light and rain can fall unimpeded to
the ground. Only then do young trees have a chance to grow
tall. With beeches, the most common native tree in ancient
forests in Germany, this handover from one generation to

the next happens every three hundred to four hundred years. Genetic modifications of characteristics related to climate change take just as long—too long.

Mutations are not the only way organisms adapt to changing environmental conditions, however, as we know from our own experience. Humans have undergone barely any genetic change in the past few thousand years, and yet in a relatively short time we have radically changed how we live. Our forebears gathered experiences and learned to deal with change by altering not only their genes, but also their behavior. Behavioral adaptations were the only way our species could settle both the icy North and blazing hot savannahs. The key to survival for long-lived species, therefore, is to learn and pass knowledge down to the next generation. And this is clearly what trees do—as you can see for yourself the next time a hot summer comes along.

THE ANCIENT BEECH FORESTS growing near my forest academy had proved to be amazingly resilient during the hot summers of 2018 and 2019. In the surrounding commercial plantations, spruce and pines were not the only trees dying. Old deciduous trees were dropping their leaves as early as August. The situation was very different in untouched forest preserves. Here, the trees' mighty crowns cast constant shade, and even after months without rain, the forests were still pleasantly damp and cool.

It was the third summer of drought—2020—when the picture changed. Up until July, it looked as though 2020 was going to be a repeat of the previous two years. The heat wave that hit in August, however, proved to be too much. Forested slopes turned completely yellow-brown, and within three days a huge number of leaves began to fall. It feels somewhat spooky to walk under trees in midsummer with millions upon millions of leaves fluttering down around you. That was when I first began to fear for the future of the beech forests.

The trees growing on north-facing slopes were most affected. These locations offer particularly good conditions for forests but were the very places where the signs of drought were most noticeable. How could this be? On north-facing slopes, the sun shines on the ground for fewer hours per day than it does on south-facing slopes for one simple reason: the ground is shaded not only by the trees but also by the mountain itself. More shade means lower air temperatures. Water also evaporates more slowly in these conditions. Shady and cool. These are exactly the conditions beeches and oaks like best. You notice the difference in how they large grow. Trees growing on north-facing slopes can be twice the size of trees on south-facing slopes, where heat and lack of moisture slow down photosynthesis. In short: north-facing slopes are a paradise for trees. Or at least they were up until now.

South-facing slopes, in contrast, have always been places of scarcity when it comes to meeting trees' needs. The slope is oriented toward the sun like an enormous solar panel, catching the full heat of the sun's rays all day long. Rain evaporates much more quickly both from the canopy and from the ground. On hot summer days, beeches and oaks on this side of the mountain run out of steam noticeably sooner. They end up photosynthesizing and producing sugar for significantly fewer days than their colleagues on the north-facing slope.

You could say that on south-facing slopes the trees are already experiencing the kinds of temperatures and rates of evaporation that trees growing on north-facing slopes will begin to experience for the first time as the climate changes. And yet here on the south-facing slope, the trees' stress as exhibited in the browning of their leaves was far less pronounced. It's not that the trees on the south-facing slope were unaffected by the drought. It's just that they switched to emergency mode in time because they already knew how to be frugal: they had started to use less water and had more or less shut down.

On the north-facing slope, however, the hot August days hit trees that clearly did not see trouble coming. Even in the middle of the 2019 drought, the moisture in the shaded ground had been enough—and it was still enough until July 2020. But then, in an instant, the trees used up the last reserves of water. It happened instantaneously because on a hot summer's day a mature beech transpires up to 130 gallons (500 liters) of water. If no rain falls to replenish supplies, any tree that doesn't apply the brakes soon enough suddenly finds nothing but dust at its feet. Its roots register the abrupt onset of drought, but by then it's too late for the tree to switch strategies. It's no longer possible for it to dial down its use of precious moisture. The only option left is to apply the emergency brake.

The trees on the north-facing slope hit the emergency brake that August. In a state of panic, they shed a massive number of leaves to reduce transpiration. The speed of their reaction was enough to tell me how dire the situation was. A tree that sheds most of its leaves in just three days is moving at top speed. Compare this to how trees shed their leaves in fall. That process starts with the gradual removal of chlorophyll, the green pigment that makes photosynthesis possible. The chlorophyll is broken down and stored in branches, in the trunk, and in the trees' roots so the pigment doesn't have to be manufactured from scratch every year. As the green is removed, the yellow pigments in the leaves are revealed. Once all the important nutrients have been extracted from the leaves, the tree grows a corky abscission layer and the leaves fall to the ground. The whole process happens at a leisurely pace over weeks, finally coming to an end in November.

The emergency jettisoning of leaves in August 2020, in contrast, was a reaction executed in a moment of complete panic. At first the beeches attempted to follow the same procedure as in fall, following the script as best they could. However, they quickly realized they were being too slow and were still losing too much water. Any tree that doesn't get ahead of the curve in these circumstances will dry out and die.

And so the beeches upped the tempo and dropped not only brown (that is to say, empty) leaves, but also yellow and even green ones. Beeches that drop green leaves have gone straight to high alert. Any tree that throws away valuable nutrients instead of retrieving them from the leaves (as they do in fall) is living dangerously. In spring, they are going to need the last of their reserves to wake up from their winter break and grow new leaves. They will have no energy left over to defend themselves should they then be struck by disease or the next drought, and they will die. Beeches, therefore, discard green leaves only in the direst of circumstances.

Despite the trees' haste, I could still discern some order in the chaos on the north-facing slope. The first leaves to be discarded were those high up on the crown, and then those on successively lower branches. This strategy worked out well for most of the trees. They were in luck when the wind took a turn to the north and moist air streamed into the Eifel mountains. Clouds climbed the slopes, releasing huge amounts of rain that quenched the parched trees' thirst. They stopped dropping leaves and even delayed leaf drop in fall. This behavior is not unusual. Instead of dropping their leaves in October as they usually do, hungry trees often hang on to their remaining leaves until November. That way, they can snack on a bit more sugar and stockpile some for the coming winter.

From a distance, the situation in forests stressed by drought often looks more disastrous than it really is. The outer leaves on the crowns are the first to turn from green to brown, which makes stands of beeches and oaks look pretty dismal when viewed from a long way off. But when you're walking in such forests, they often look surprisingly vibrant. As you wander under the canopy, mostly what you see are the leaves of the inner crowns, which are still lush and green. The trees are on high alert only when all their leaves are strewn on the ground by August.

Most of the trees on the north-facing slope in Wershofen will survive this shock to their systems. The most important thing is that they have now learned to ration water better. For the rest of their lives, they will slow down and drink only in moderation, and come spring, they will not completely exhaust the rain stored in the ground over winter. This change in behavior is measurable. It manifests itself as slower growth in the diameter of the trunk. After this traumatic experience, even in the absence of drought in the future, the trees will faithfully follow this new strategy—just in case...

WHEN WE CHANGE our behavior based on new experiences, we call it learning—and learning is the most important survival strategy of life-forms that live for a long time.

Plants can learn in significantly more complex ways than the trees' reaction to drought. To check this out, let's take a step back and consider peas. These legumes have the unbeatable benefit for researchers that they are much easier to manage in the laboratory than beeches or oaks. In this carefully curated world, what the tiny plants reveal is nothing short of astonishing.

Monica Gagliano, a biologist from Sydney, Australia, trains peas the way you might train a dog. You have no doubt heard of the historic experiments of the Russian physiologist Ivan Pavlov. He was researching how dogs behave. When he offered them food, they began to salivate; when he rang a bell, nothing happened. Then he began ringing the bell before he fed them. Soon the sound of the bell alone caused the dogs' saliva to flow, even if there was nothing to eat. This process is known as conditioning—a connection is made between two completely different stimuli and both now elicit the same reaction. And as it turns out, you can condition peas as well.

To do this, Gagliano kept the peas in the dark until they got hungry. Then, every so often, she shone a blue light on the little plants. Light is the energy plants use to photosynthesize, and by now the peas were getting really peckish. They quickly turned their leaves toward the light—something you've probably noticed your houseplants do as well. That in itself was nothing out of the ordinary, with the possible exception that when the peas found themselves in the dark once again, they returned their leaves to a neutral position.

The scientist now added a puff of air right before she turned on the light. In the final step of the experiment, the peas received a puff of air in the dark without the light being turned on afterward. And lo and behold, the plants oriented their leaves to the puff of air, clearly expecting light would soon follow from this same direction. They associated the puff of air with the light even though this stimulus had nothing to do with photosynthesis. To put it another way: peas are able to make connections.

Gagliano believes many plants probably possess this ability.[2] The results of her experiments demonstrate that our fellow beings clad in green are capable of much more complex levels of learning than we have hitherto suspected. It follows that their ability to adapt to change must also be more developed than we thought. And with that, let's return to the topic of trees.

PARTICULARLY IMPRESSIVE TREES growing in what is now a wildlife park near Ivenack in the German state of Mecklenburg-Vorpommern demonstrate that trees can continue to learn for extremely long periods of time. The pedunculate oaks there have short, thick trunks and mighty, gnarled branches. They are estimated to be between five hundred and one thousand years old, which makes them among the oldest trees in Germany. The most imposing trunk has a diameter of 11.45 feet (3.49 meters) and a volume of 6,357 cubic feet (180 cubic meters)—360 times the average for a tree in this country.[3]

Foresters generally think of old trees as frail. They assign them a low value because the interior wood of ancient trees has often been rotted away by fungi, which means the trunks can no longer be sent to sawmills for processing. The accepted wisdom among state foresters is that ancient trees are the ones least able to defend themselves against heat and drought. It makes sense, therefore, to cut them down before they rot anymore and to replace them with young, vigorous trees. But this narrative is simply a public-relations ploy that allows foresters to cut down large, valuable trees without having to worry about public protest. And that is why you can no longer find ancient trees in German forests. The only places where they still grow are in parks—where trees are not managed for commercial gain and people love them for their own sake.

The oaks of Ivenack had a hard life even before climate change. After all, you can't have a genuine forest microclimate in a cultivated space like this. You would expect them to live shorter lives than their colleagues in true forests. And yet they are the record holders among our native oaks—and their longevity is connected with the way they learn.

Researchers took a close look at the oldest oak. CT scans are a good, noninvasive way to look inside people, and this is true for trees as well. Scans revealed that beneath its thin outer layer of wood, the colossus was rotten and hollow. Even though the trunk was nearly 11.5 feet (3.5 meters) across, the outer layer of reasonably healthy wood was no more than 2.5 to 20 inches (6 to 50 centimeters) thick. In places, it was no longer capable of bearing any weight. The tree had to use what wood remained to withstand storms, transport water up into its crown, and carry food back down to its roots. Is it any wonder that in the dry year of 2018 the oak looked ragged and people were seriously concerned about it? In addition, these old warrior oaks grow in a park where large herds of mouflon sheep and fallow deer deposit copious quantities of droppings on the ground. This leads to overfertilization with nitrogen, something trees do not enjoy at all.[4]

In 2020, worried by three consecutive summers of drought, a team led by Andreas Roloff decided to see how the oldest oak was doing. It quickly became clear that the oak was doing well. According to Roloff, a review of the foliage and branches led the team to conclude that the old-timer was in prime condition for a tree its age.

To get a closer and more detailed look, the scientists threw a rope over a branch in the crown and gathered samples. To their surprise, they found a number of leaves on the tree's new growth that belonged to sessile oaks, that is to say, to a completely different species. And that was not all: along with acorns that also looked as though they belonged to sessile oaks, they discovered leaves usually found on Pyrenean oaks. Different species of oak, all together on a single tree?

AN INTRIGUING THEORY has been drifting through the ranks of professional foresters that suggests there is no such thing as a sessile oak or a pedunculate oak. Rather, there is just one species that looks different depending on where it grows.

The acorns of pedunculate oaks have long stalks (a peduncle is a stalk, hence the name), and their leaves also look slightly different from those of sessile oaks. Location, however, is the main factor that distinguishes the two species. Whereas sessile oaks occupy dry spots on hills and mountains, pedunculate oaks often grow in bottomland riparian forests, where they tolerate months of flooding. Nature has, therefore, equipped them with what they need to feel more at home at lower elevations.

The idea of two separate species has held sway in professional forestry circles up until now. And yet the diagnostic differences in the leaves and acorns are not quite so distinct out in the forest itself. Both species appear to happily interbreed, and their offspring display all kinds of intermediate characteristics. Add to this the recent investigation of the Ivenack oaks, and the theory that has been floating around in the background now seems more plausible. Perhaps we are not talking about two species after all. Perhaps we are talking about only one species that takes on different forms as it adapts to the prevailing climate.

Genetic analysis has shown that the ancestors of the ancient trees in Ivenack migrated back north from Spain after the last ice age. As it gets warmer and drier in Germany once again (as it once did in their Spanish home), it's possible they will adapt to these conditions and this adaptation will be reflected in the varying shapes of their leaves. The fact the oaks recovered after 2018 despite the two extremely dry years that followed (in 2019 and 2020) supports this idea.[5] To put another way: perhaps the old trees remember their ancestral homeland.

Another possibility is that we are witnessing the rise of a new species. *Witnessing* is meant in relative terms here, as it can take thousands of years for a new species to arise. Perhaps Germany's native oak is dividing into two new species: the pedunculate oak and the sessile oak. That sounds a little unlikely, as hybrids are popping up the length and breadth of the country. Oaks are wind-pollinated and their pollen can travel many miles to the next tree. This results in constant mixing. How does a new species get established if the up-and-coming generations are always comingling?

Something similar is going on with one of Germany's native birds, and the obstacles are much the same. I'm talking about carrion crows, which are probably also in the process of creating a new species. They, too, can fly long distances, and when they do so, they mix with crows in other parts of the continent. Nevertheless, a new color variant is emerging: the hooded crow. Genetic analysis has shown that carrion crows and hooded crows belong to the same species and interbreed all over the place. Despite this, one tends to be more prevalent than the other, depending on the territory. For example, we don't have any hooded crows in the local forests around Wershofen, but east of the Elbe River, you mostly see only hooded crows and no carrion crows at all. Although the crows of both color variants can pair up, they hardly ever do. That is due to a phenomenon that my wife and I have also observed with our chickens and even with our goats: individuals are more strongly attracted to other individuals with similar coloration. And that is how hooded crows maintain distinct populations—and in the future will likely develop into their own species.

This attraction naturally doesn't work with oaks. Pollen, after all, cannot choose which female flower it would like to land on and which it would rather avoid. The explanation we are left with is that the oaks are adapting to the locations where they are growing and to a changing climate, and this adaptation is being manifested in changes in the appearance of the trees' leaves and fruit. The theory of two separate species doesn't sound too plausible to me.

THE INVESTIGATIONS INTO the Ivenack oaks have shed light on a completely different topic: even the oldest trees are capable of adapting to different environmental conditions. As you may already know from my book *The Hidden Life of Trees*, trees are capable of learning and can store the knowledge they have acquired for a long time. Trees that have been learning for a thousand years should know how to respond more effectively to summer drought than freshly planted young seedlings. The research results, therefore, are also a plea for us to finally let our forest trees grow old once again.

Lifelong learners accumulate a lot of knowledge. We store our knowledge in books and computers or, as in days gone by, we pass it on by word of mouth. But how does this work with trees? Do all their experiences die with them when their lives come to an end? That's what we thought for a long time, until a young scientific discipline set out to answer this question and determined that trees, too, pass their wisdom on to the next generation.

3

Seeds of
Wisdom

THERE'S A MAD RUSH in the forest these days—or, more precisely, in forestry circles. How can we prepare forests for climate change, for heat and periods of drought? Although trees can learn, when it comes to adapting genetically, they are, unfortunately, extremely slow. Mutations—changes in genes and the attributes associated with them—are not expressed until the next generation. In a naturally growing forest, where the new generation of trees might spend hundreds of years in the shade of mother trees, these mutations don't take effect until after the mother tree has died of old age. Depending on the species, this can take as long as six hundred years. In times of rapidly advancing climate change, that is, of course, far too slow. Many animals have a clear advantage here. Take hares. They reproduce so rapidly the females can become pregnant again while they are still pregnant. That's how they can have from three to four litters a year with the corresponding opportunities for genetic variation and adaptation.

Mutations, however, are not goal-oriented, which means they are not a particularly effective way of making changes. They are merely errors reading the genetic code during reproduction. Most mutations, therefore, never amount to anything—and even when they do, they might lead in completely the wrong direction. It could take thousands of years for a process like this to accidentally produce better adapted trees. Wouldn't it be better to take chance out of the equation and hurry the process along? That's what people do. We pass our experiences down by word of mouth or in writing. Instead of waiting for mutations to help them adapt, members of the next generation then change how they live their lives. Trees have no writing, at least not in the traditional sense. And yet they write a kind of summary of their experiences for their offspring, a summary they encode in their DNA. Before we take a closer look at how trees do this, let's step back into a time shortly after the Second World War.

JUST A FEW DECADES AGO, prevailing scientific opinion held that genetic changes are possible only through mutations and not through experience, meaning that the only way experiences could be transmitted to the next generation would be through words or actions. But this opinion changed because of the Second World War. In the winter of 1944–45, many Dutch people were starving because of food shortages caused by the German occupation. Pregnant women evidently passed this experience on to their unborn children, as their metabolisms were programmed for food scarcity. Food abundance in the postwar years caused these children to experience more health problems later in life. Compared with the average Dutch person, they were more likely to be overweight and to suffer from other so-called diseases of civilization.[1]

Every single cell in our bodies shows it is not genes alone that shape what we look like and how we function. Each cell contains an identical blueprint for a complete human being, all tightly wound inside it. If you were to stretch out a strand of DNA, it would be about 6 feet (2 meters) long. It contains an abundance of information at a molecular level, only some of which is used at any one location in the body. The cells in your hand develop differently from those in your brain. How does your body control what happens during growth and healing so that only the type of cells needed in each location are formed?

This is where epigenetics come into play—in other words, the processes that determine which genes are turned on or turned off. You can imagine DNA as an encyclopedia that contains all there is to know about how a human body is constructed and how it works. Epigenetic processes are a bit like bookmarks that help ensure the encyclopedia opens only to those pages that need to be read. Inserting bookmarks happens with the help of methyl groups that attach to genetic material and modify it. These modifications are influenced by an individual's life experiences, as the example of that winter of starvation in the Netherlands demonstrates.

Trees have the capacity to pass along experiences too—as scientists at the Technical University of Munich revealed when they investigated an ancient poplar. The 330-year-old tree has had to constantly adapt to changes in its environment, such as droughts or fluctuations in temperature, and you can clearly see this in its genes. But how could the scientists tell that the genes of this ancient tree had changed? It was actually quite easy. They looked at leaves widely spaced from each other along a branch. Every year, the branch grows longer and older. The oldest section is close to the trunk, where the branch originally began to grow. The youngest section is at the tip. If the poplar had indeed learned over the centuries and epigenetics had altered its genes, one would expect to find the greatest changes at the branch tips.

And this is exactly what the researchers found. The farther apart the leaves were on the branch, the greater the differences in their "bookmarks." In the poplar under investigation, the changes happened up to ten thousand times faster than they would have had they been caused by mutations passed down from one generation to the next. Moreover, we know that most of the time, trees pass the sum of all these innovations (or experiences) on not only to their immediate offspring but also to the many generations that follow.[2] And as trees reproduce annually, they can produce offspring with newly adapted characteristics every year.

How, though, do you find out if the young trees have indeed learned anything from their parents? The procedure takes time but is not complicated. Scientists from the Swiss Federal Institute for Forest, Snow and Landscape Research had been watering plots of Scotch pine for various experiments since 2003. The trees wanted for nothing; you could say they were spoiled. In 2013, the scientists stopped watering some of the plots. Three years after the watering stopped, they gathered seeds from both the pampered pines and those deprived of water and planted the seeds in a greenhouse. Lo and behold, the seedlings from mother trees that had been watered regularly were far less able to handle drought than the offspring of pines deprived of extra water. This was one of the first pieces of evidence that trees pass their knowledge on to the next generation.[3]

A similar experiment had trees traveling somewhat far-ther afield. Spruce from Austria were transplanted to rugged Norway. When the trees matured, they produced offspring. And here, too, you could see the effects of learning in the next generation. The seedlings showed the same frost tolerance as their Norwegian colleagues. And learning works in the other direction as well. Norwegian spruce transplanted far-ther south adapted to the warmer climate and their offspring were not as frost-tolerant as the mother trees.[4]

The idea that adaptation in trees takes forever because of their long lives and the correspondingly long time between generations proved to be incorrect. As tree parents continue to learn until their dying breath, the seeds they produce are equipped with all their latest survival strategies. Tree off-spring, therefore, do not need to start from scratch and learn from all their mistakes—thanks to epigenetics. This means that the grand old age of mother trees is not a disadvantage. On the contrary, it is a great advantage as it means older is wiser with better-adapted offspring. Young trees profit from the centuries-long experience of their parents—and this brings us back to the reproduction-happy hares. A hare lives ten years at most and therefore can pass relatively few skills along epigenetically. This time, it is the trees that clearly hold the advantage.

The easiest way to ascertain what trees have learned in their lifetimes is to look at the outermost branch tips. The knowledge of a lifetime is concentrated in the most recent growth on an old tree. In the case of the Ivenack oaks, this is an impressive one thousand years' worth of learning. The surprising differences in the leaves that indicate the change from pedunculate oak (when it's moister) to sessile oak (when it's drier) are mostly found in the upper branches in the crown—the youngest branches on the tree. It would be fascinating to know whether the acorns on these branches grow into young trees that are better adapted to drought than trees that sprout from acorns growing on older branches. Recent research suggests this may well be the case, and that would be further proof that trees can adapt to climate change much faster than we have suspected up until now. Whether the speed of adaptation is enough depends, of course, on the extent to which we keep increasing the tempo of change through unchecked environmental destruction.

Beeches and oaks like it to be cool and moist, and therefore increasing summer drought is worrying; however, it might not be the biggest problem trees face—as you will see.

4

Filling Up
in Winter

DAMP ENOUGH AND NO WIND. In weather like this, I don't worry about trees. In winter storms I keep an anxious eye on their crowns as they creak and bend—and hope not too many trees will blow over. When it's hot and dry in summer, I think of spruce already dealing with thirst also being sought out by bark beetles. When a storm hits in the summer, finally bringing much-needed rain, I'm concerned. Deciduous trees are designed to withstand winter storms. They drop their leaves in advance to reduce the area exposed to the wind. Thanks to their stripped-down branches, they don't topple over as easily as needle-clad conifers. When summer storms gather, ushered in by short but powerful gusts of wind, the situation is completely different. Beeches and oaks in full leaf bend deeply in the unexpected squalls. When deciduous trees fall or break, they usually do so in this kind of weather. And so, you see, foresters can almost always find a reason to worry.

A team from the Swiss Federal Institute of Technology in Zurich (ETH Zurich) went some way to calming my concerns about summer drought, at least. The researchers investigated 182 forests in Switzerland to see which season produced the water that beeches, oaks, and spruce drink during the summer. Your immediate answer would probably be: "Summer, of course. When else?" The surprising answer, however, is that they mostly drink water left over from the winter. Therefore, summer rainfall is less important than how often it rains in winter. But before we think through the implications of this finding, another question needs to be answered. How did the researchers discover this in the first place?

They started by looking at the differences between winter rains and summer rains. To do this, the scientists probed the ground with lysimeters. A lysimeter is a piece of equipment that allows researchers to take samples of water up to 50 inches (120 centimeters) underground. Rains that fall in winter have a different chemical signature from rains that fall in summer, and the water they produce is bound up in different soil structures. But how did the researchers know which moisture the trees were using? That's easy—they checked out the signature of the water in the branches in the trees' crowns. Well, maybe not that easy, because technicians needed to hang from a helicopter to cut branch samples from the crown. The laboratory results showed that even in summer, beeches and oaks are drinking water from winter rains. Spruce, in contrast, are more dependent on year-round rain.

Now, you might think the trees are drinking more winter rainwater because it rains less in summer. But that was not the case in the locations where the Swiss scientists carried out their investigations—about 58 percent of the total annual rainfall there falls in the summer. Also, if the trees were simply drinking whatever water was available, you would expect there to be no differences between tree species.

The scientists explained the differences between the deciduous trees and the spruce they studied by pointing out that oaks and beeches prefer to suck moisture from fine pores deep in the ground, whereas spruce drink more water from larger pores—in the same part of the forest. That way, tree species do not compete as directly as some foresters believe, even though they root to similar depths. This also helps explain how winter precipitation and summer precipitation remain separated in the ground. Whereas summer precipitation is immediately taken up by plants and transpired, precipitation that falls in winter, when the trees are all sleeping and using barely any water at all, is left to gradually seep into the smallest spaces in the soil until the pores are completely saturated.[1] Depending on the soil, up to 50 gallons (200 liters) of water can be absorbed and stored per square yard (meter) of forest floor.[2]

This newfound knowledge makes two issues clearer. If we want to know how the trees native to central Europe are doing, we should be looking more closely at precipitation levels in winter. These are inevitably going to decrease, if only because winters keep getting shorter thanks to climate change. According to the German Environment Agency, winter is now fourteen days shorter than it was in 1961.[3]

The other challenge trees face is the destruction of the spaces for storing water when timber-harvesting machines that weigh as much as 70 metric tons drive over the soil. Soil is delicate and reacts to pressure like a sponge that is being squeezed; however, unlike a sponge, soil does not bounce back—ever. Hardly any water seeps down through the tire tracks, and there isn't enough water in the ground for trees to fully recharge. Under normal circumstances, when the ground is undamaged, the supply of water from winter clearly provides a comfortable cushion, even in times of drought, acting as a reservoir the trees can access all summer long.

We can now look at fall leaf drop completely differently. Up until now, we thought the main reason trees shed their leaves was to avoid heavy loads such as wet snow on their branches. When snow trapped by leaves absorbs water, it gets enormously heavy. This weight soon overwhelms the tree, thick branches break off, and sometimes the whole tree falls over. A tree with bare branches is also better equipped to withstand storms because it offers no large areas for hefty squalls to batter.

The results of recent scientific studies, however, show that interception might be another important reason for leaf drop. Interception has to do with the amount of rainfall that gets trapped in the trees' leafy crowns—and it's a lot. This moisture evaporates directly off the leaves back into the air. It is lost to the trees, which means in summer only heavy rainfalls quench their thirst. For their downtime in winter, in contrast, it makes sense for them to "undress," as they're not using their leaves then anyway. That way, raindrops can reach the ground directly without leaves getting in the way. In summer, the forest is the vegetated area that loses most water through evaporation. Per square yard (meter) of ground, there are up to 32 square yards (27 square meters) of leaf or needle surface up in their crowns.[4] Only when these are dripping wet does additional rain fall to the ground. Trapped water evaporates back into the air without ever reaching the ground.

While all trees hold back a similar amount of rainfall in their crowns in summer, there are big differences in winter. It's easy to see how deciduous forests allow more rainfall to reach the ground. Beeches, oaks, and other deciduous trees allow the rain to rush almost without hindrance to the ground. Water that lands on branches, which angle upward to the sky, is directed as if in a funnel and shoots in small torrents over the bark and down to the roots. The situation is very different with evergreen pines and spruce. For them, the seasons are much the same. In a coniferous forest in winter, between 30 and 40 percent of the rainfall is still captured by their crowns. This is in contrast to the deciduous trees standing naked at this time of year, where the amount of water captured decreases to under 8 percent.[5] You might wonder what happens to rain caught up in the crowns. Although it is lost to this particular forest when it evaporates from leaves and needles, the rising water vapor forms new rain clouds that soak forests elsewhere. And so, even though individual forests are affected, the forest ecosystem as a whole is fine.

For the trees, all that matters is the amount of moisture that ends up in the ground where their roots can reach it. In coniferous forests, this is about one-third less than the amount the rain clouds drop on their crowns. Why, then, do spruce and pines do something as "foolish" as keeping needles on their branches even in winter? Water is, after all, the number one elixir of life. The reason lies in the trees' homeland way up in the band of coniferous forest that circles the northern part of our planet, where summers are short and winters are long. In these latitudes, any tree that grows leaves every spring and discards them in fall has hardly any time to produce sugar through photosynthesis. In this part of the world, it's best to remain prepared so you can get going as soon as temperatures allow.

To finally reach the trees' roots, rainwater must penetrate the often thick layers of discarded leaves and needles that accumulate on the forest floor. It's just as well the leaves of trees native to central Europe decompose quickly. The army of ground-dwelling organisms in this part of the world throws itself enthusiastically upon the fallen biomass and consumes it at a breathtaking rate. The mass of leaves and woody debris digested in Germany's native forests can amount to as much as 5 tons per 2.5 acres (5 metric tons per hectare) per year, with fallen leaves over this same area numbering in the millions. A single beech can discard half a million leaves that pile up in a layer up to 4 inches (10 centimeters) thick at its feet.[6] Depending on the quality of the soil, this can be decomposed and turned into crumbly humus within one to three years. This humus layer is one of the most important reservoirs of water on the forest floor. In essence, leaves are repurposed as water tanks for the trees.

This system doesn't work nearly as well in monocultures of spruce and pine. It's generally thought the problem arises because the ground-dwelling organisms find the acidic needles distasteful, but I think it's more likely that our native ground-dwelling organisms simply aren't as effective when confronted with needles full of terpenes and resin. Rain that manages to reach the ground through the thick crowns of spruce and pines now meets with its next obstacle: needles that have accumulated over the years to form a thick carpet on the forest floor. I've often noticed this carpet acts as though it's been waterproofed: after a prolonged dry period, water runs off it instead of soaking in. It's little wonder, then, that right now so many trees in Germany's conifer plantations are giving up the ghost as summer drought increases, especially as spruce are more dependent on this summer rainfall than beeches and oaks are.

THE QUESTION OF which trees are growing aboveground is becoming more pressing for our groundwater supplies.

The only water that penetrates this far down is water left over from many different processes. Before a single drop of water seeps underground, large amounts have evaporated from treetops, flowed off impervious surfaces, and been stored in humus and soil. And this doesn't include the water a mature tree siphons off for itself. It can drink up to 130 gallons (500 liters) on a hot summer's day. Only crumbs (make that drops) fall from the trees' table and the leftovers vary greatly. Naturally growing beech forests are exceedingly generous compared with pine plantations, allowing three to five times more moisture to trickle down than their needle-bearing colleagues.[7]

LARCHES ARE THE EXCEPTION among conifers. They are at home in the mountainous areas of Europe and the only native conifers that discard their green finery in fall along with the deciduous trees. They are often grown in plantations alongside spruce and pines, where they are, unfortunately, equally out of place. Even so, larches manage a little better. For one thing, their bare crowns in winter offer as little surface area for storms to batter as those of deciduous trees. For another, from November to April, like beeches and oaks, they offer little obstruction to rain beating down on the ground. I think it's no coincidence that larches naturally grow in moister regions and require more water than, for example, pines. What better idea, then, than for them to adopt a strategy like the one used by deciduous trees?

As fall approaches and the leaves gradually turn, you can evaluate trees' fitness for yourself. Just take a look to see what color they are turning, and, at least with some species, this will tell you exactly how they are doing.

5

Red Flags
for Aphids

IN OCTOBER 2020, I noticed something unusual. The colors of the fall leaves were not as bright as usual. I could see green, yellow, and brown, just like every year. Yellow, especially, glowed as cheerfully as ever, bathed in sunlight streaming through gaps in the clouds. And yet the cherry trees and a mighty pear tree in our horse pasture usually had more to offer: a fireworks display ranging from orange to deep red. The cherries, in particular, usually began to color their leaves brightly by the end of August. The cherry trees are usually full of sugar by then, and so they shut things down. In years that are both moist and warm, they seem to lay in sugar reserves more quickly than other trees, and when their storage spaces are full, it doesn't make any sense for them to keep on photosynthesizing.

In the summer of 2020, however, the cherry trees looked different. I found no trace of the vibrant colors that indicated they were full of sugar. All I could see was the occasional browning leaf, a sign the trees had been stressed by lack of water, just like all the trees in the surrounding forest. I wasn't surprised when the cherry trees turned from green to yellow not at the end of August, which is what they normally did, but at the end of October, when all the other deciduous trees were turning color. Clearly, they had waited until then to withdraw reserves of nutrients and minerals from their leaves.

Trees do not actively add yellow to their leaves. The color is revealed when the trees break down chlorophyll, which is green, and gradually draw it back into their branches, trunk, and roots as they prepare to hibernate for winter. The component parts of chlorophyll remain stored in their tissue until they travel back into the new leaves the following spring. When the green disappears from the leaves in fall, yellow carotenoids are revealed. They are there all the time, but the green pigment keeps them hidden.

Red is completely different. A tree creates this color and then actively pumps it into its leaves. Red is like a car driving the wrong way down a one-way street. Constructing red costs trees time and energy, and researchers are still trying to puzzle out why they go to all this trouble at a time when the threat of being surprised by the early onset of winter increases by the day. Instead of busily sending another pigment out into the fray, trees would surely be better off focusing all their energy on bringing the last remnants of useful substances back into the safety of their trunks and branches as quickly as possible. When the first massive cold snap comes, beeches, oaks, and other deciduous trees have no choice but to shut down for the winter. Any substances not retrieved by then are lost.

One common explanation goes like this: trees produce something akin to sunscreen for their leaves—a bit like the process that gradually darkens our skin when it's exposed to ultraviolet rays. But why would leaves need to be protected against sunlight just before they are discarded? Researchers propose that leaves are particularly fragile when chlorophyll is being deconstructed and withdrawn.[1] The leaf cells are still alive, and the tree must remove residual materials from them as quickly as possible if it is to save and store them for another year—the tree couldn't do that if the living tissue was damaged.

Another explanation, however, sounds at least as plausible. The red pigment warns off insects looking to suck juices from leaves. Trees with red leaves are advertising their fitness. It's as though they're saying "Check this out, parasites! Even in fall, I'm strong enough to go that extra mile and color my leaves red. Don't even think of landing here because next year I'll poison you!" According to this explanation, trees that color their leaves red are showing off. But things are not that simple, because the insects—aphids, for example—are an important part of this conversation. Red can't impress them because they don't have receptors in their eyes for that color, which means they can't see it. And yet there's evidence to suggest that red plays a role in the choices they make. More on that later.

JUST LIKE THE TREES, insects are also preparing for winter. For many of them, this is easy: they die. But before they do so, the females produce eggs one last time. The insect mothers lay their eggs in cracks in the bark of suitable trees so that when their offspring hatch the following spring, they are not too far from the nearest source of food. The trees defend themselves from attacks by parasites by storing toxic substances in their needles and leaves—but they can do this only if they are healthy. You have only to look at spruce and pines in large plantations in Germany these days to see how easy it is for insects to attack sickly trees. Bark beetles can literally smell when conifers become so stressed they can no longer defend themselves when the beetles try to bore into them.

The relationship between aphids and apple trees is similar to the one between bark beetles and spruce. You might recognize this scenario from your own garden. As soon as the first delicate leaves appear on the apple trees in spring, a number of them start to curl like claws. A quick glance at the underside reveals the reason. Armies of aphids are pouncing on the tender new growth, sticking their snouts into it and sucking out the sweet juices. A heavy infestation causes new shoots to wither away, which stunts the growth of the tree. Damaged branches grow hardly any leaves, which considerably weakens the tree. The uninvited guests can also carry viral, fungal, and bacterial diseases. In short, aphids are a huge pain if you are an apple tree. Wouldn't it be nice if the bothersome little creatures gave your crown a wide berth?

And that's exactly what a healthy apple tree gets the aphids to do when it colors its leaves red in the fall. You can see something similar happening with many other deciduous trees that color their leaves red at this time of year. Long-running scientific studies show that basically all species of trees with red fall foliage are targeted by a particularly large number of species of aphids, which they have to fight off. It seems to be a case of coevolution—of eating and being eaten.

As I just mentioned, aphids cannot see the color red. And yet there are experimental results that prove aphids are less likely to attack trees that add red to their foliage. Moreover, aphid offspring on these trees are not as healthy, in stark contrast to aphids on trees of the same species that carry only yellow leaves in fall.[2]

It follows, then, that trees cannot use red as a warning, even though this would be a lovely analogy to our own senses. The answer is there, however, if we take a moment to look at the world through the aphids' eyes. In fall, they search for trees that will give their offspring the best start in life, and green and yellow foliage jumps out at them. They are attracted by these signals, which trigger them to lay their eggs on the trees' trunks and branches. Red is not a warning after all, but camouflage. What stands out to us disappears into a nondescript mixture of blue and green when viewed by aphids.[3]

Marco Archetti of Harvard University wanted to take a closer look at the aphid–tree relationship. To do this, he investigated the health of aphids on apple trees. Apple trees are good subjects for study because scientists can observe and compare both wild trees and a wide range of cultivated varieties. The wild trees have been able to adapt to the interplay with aphids over thousands of years and have developed their own strategies; the cultivated varieties haven't had this luxury.

We humans, with our focus on yields, breed trees that produce large, beautiful, tasty apples. Evolutionary pressures on these trees no longer come from aphids but from us. And therefore, the trees adapt to our wishes—our selection criteria—because these are the main factors affecting their survival in cultivated landscapes. In the process, unfortunately, we ignore other characteristics, especially those whose significance we have, until now, known nothing about. In the past, red as a defensive color against aphids played no role in apple breeding. And why should it have when gardeners had no idea what was going on? As apples have been bred for thousands of years, in many cultivated varieties red coloring in fall has disappeared.

To underscore the connection, Archetti evaluated the aphids' survival rate in spring. In apple trees with green foliage in fall, it was about 61 percent. In those with yellower foliage, it dropped to 55 percent. And in trees with reddish fall coloring, it sank to 29 percent. If red coloring is a good defensive strategy, then why don't all apple trees employ it? Archetti believes there is a further explanation beyond breeding programs with their focus on other characteristics: how susceptible trees are to diseases such as the dreaded fireblight, which is transmitted by aphids. Species that are particularly susceptible must defend themselves with corresponding vigor against the carriers of this disease, whereas robust species are more able to tolerate aphid attacks. And this was exactly what Archetti's field studies of apple trees in North America showed. Cultivated varieties that still colored their leaves red in fall despite being products of breeding programs were those varieties most affected by fireblight.[4]

NOW LET'S RETURN TO October 2020. After a long, hard summer, there was barely any red to be seen in the trees in many areas of Germany. Cherries, apple trees, and bushes such wild sloes were turning from green to yellow, but then few were making it beyond pale orange. When you know creating red pigment is an active process that demands resources, this comes as no surprise. It's important for trees to mount a defense against aphid attacks so they can come through spring fit and healthy; however, before that, they need sufficient resources to survive the winter.

The deciduous trees in Germany were in a similar situation to brown bears that haven't caught enough salmon before taking their winter break and whose layer of fat is therefore too thin for hibernating. A tree that is worried it won't survive the coming winter because it has stored only a limited number of calories won't invest many of them in changing the color of its leaves. Mounting a defense against aphids is a problem it won't have to face until after its new leaves grow—and the tree will be producing sugar from the first moment it begins to leaf out. Even if the uninvited guests start to siphon off some of this sugar right away, the tree's chances of survival rise with every succeeding day.

A NEW STUDY from Switzerland points to the opposite problem: filling up with sugar too soon. Researchers from ETH Zurich have shown that deciduous trees are changing their behavior in response to climate change: they are discarding their leaves sooner. Scientists had previously expected that trees would delay leaf drop by two to three weeks as the climate changed and fall weather became milder. A team led by Deborah Zani discovered the opposite is happening. She predicts that in the coming decades, trees will discard their colorful foliage from three to six days earlier. This will happen because trees will be unfurling their leaves two weeks earlier in spring because of climate change, which means the leaves age earlier as well.

I don't find aging to be a convincing explanation, because after dry summers many trees hold on to their leaves for a particularly long time. I've observed this myself on our north slope. And it makes sense. When little water is available, trees can't manufacture sugar, and so they're still hungry come October. They don't discard their solar collectors until close to the end of the month, and sometimes not until the beginning of November. Leaves should therefore have no problem lasting a few extra weeks and still doing everything they're supposed to do.

It seems to me that another result of this research is closer to the mark. Zani and her team point to restricted uptake of carbon dioxide due to limited nutrients in the soil.[5] I would express this somewhat differently. If the tree starts stocking itself with food two weeks earlier than usual in spring, it seems only logical that it would also stop filling itself up earlier at the end of the season. It has to store the sugar in its tissue, and at some point it runs out of room. Trees, after all, can't put on a layer of fat like we can and expand outward to accommodate it. When they're stuffed, it's time for them to shut down their intake of food. Although they could simply close the openings on the undersides of their leaves, why would they wait any longer to start their winter break? And so they discard all their finery a few days earlier than usual—unless there's been a summer drought.

WHILE WE ARE on the topic of fallen leaves, I was trudging through one of our ancient beech preserves during the great heat wave of August 2020 when I noticed something else that was different. A thick layer of old leaves from the fall was still carpeting the ground. In previous years I hadn't paid much attention to this aspect of the woodland. However, when the ground began to dry out much more than it used to, I started doing a little test on a regular basis to see how long the moisture in the ground would last. By the way, you can try this yourself in your own yard or forest. Move the humus layer to one side and take a bit of soil between your thumb and index finger. If you can squish it flat, there's still enough moisture in the soil. If it crumbles between your finger and thumb, it means the soil is too dry for the roots.

I wondered briefly why so many old leaves were still lying intact on the ground. Then I thought of a compost pile. The material breaks down only when the pile is moist enough inside, which makes sense because fungi and bacteria cannot get to work without water. There's a reason drying is one of the oldest ways in the world to preserve food. That was exactly what had happened to the old leaves during the long-standing drought. This process of desiccation had advantages and disadvantages for the trees. A thick layer of leaves keeps the ground from drying out so quickly during a period of drought. Unfortunately, when the rain comes, this layer also prevents light showers from reaching the ground. The raindrops moisten the leaves; only when they are completely saturated does water seep into the ground.

In the winter months, the amount of precipitation is not the only thing that matters. The trees also need the weather to get really cold. If temperatures don't drop, the trees get confused when it comes time to grow new leaves in the spring. And this becomes an added burden for beeches and oaks at a time of year when they are especially hungry.

6

Early Risers and Late Sleepers

HOW MANY OF YOU have done this? You've gone for a walk in fall and brought home a beechnut or an acorn that you've then planted in a pot and set on the windowsill to watch it grow into a small tree. It doesn't last long—because it's not exposed to winter. Trees, like many animals, must rest in the cold season. Shorter days and increasing cold trigger hibernation. Without these stimuli, trees cannot fall asleep, and instead of hibernating, they die. Potted tree seedlings, therefore, can only survive in the long term if they live outside.

But even out in nature, it's getting warmer. Winter is setting in later and ending sooner. It seems to follow that the trees' winter sleep is also getting shorter. In Germany today, April feels almost like summer. We need to change the month in the title of the famous German folk song "May Has Come, the Trees Are Leafing Out." The tender green of the trees' unfurling leaves now appears on branches weeks earlier. The German weather service reports that when you add in warmer fall days, the downtime for plants has shrunk by two weeks in recent decades.[1]

As you probably suspected, this is not good for trees. Sure, they can satisfy their hunger sooner by cranking up photosynthesis in April, but climate change has barely affected one seasonal danger that still lurks out there: a late frost. On clear nights, the thermometer regularly plunges below freezing as late as the middle of May—as it did in 2020. Most of the freshly formed leaves freeze, severely impacting the health of the trees, which must now mobilize the last of their reserves and leaf out once again. If they become sick at this time, they will have hardly any strength left to fight off fungi or bacteria.

The milder the winter, the greater the danger of leafing out too early. One January it was so warm that the cranes returned from Spain. Then, in February, winter arrived in full force, prompting the birds to fly back south. Trees, anchored in place, can't travel back and forth like this. They have to bide their time and be patient. Beeches don't get their cues from temperature alone. They also wait until the spring days last at least thirteen hours—only then do they dare unfurl new leaves. Their fear of potential late frosts is apparently greater than the hunger that gnaws at them after their period of winter rest. In Germany, the average date by which the days are this long is April 23.[2] On spring forest walks, look to see if the beeches in the woods in your neighborhood are keeping to this timetable.

But back to the need for cold. In Germany, in the absence of this stimulus, our native trees don't know that there really has been a winter between fall and spring or that another six months have passed. Perhaps trees are like us. If we wake up in the dark, we cannot tell how late it is or whether we can turn over and can go back to sleep without first checking the clock.

For beeches and maples, the temperature must drop below 39 degrees Fahrenheit (4 degrees Celsius) if their buds are to open as they should later in spring. Without this drop in temperature, the trees don't wake up at the right time from their winter sleep—it's as though they're still waiting for winter to arrive. In extreme cases, the buds on some branches never do open.[3] Contrary to the popular belief that warm winters will prompt trees to leaf out early, they can cause exactly the opposite to happen.

TREES CANNOT INFLUENCE how cold it gets in winter. In summer, however, it's a completely different story. Shimmering heat combined with long-lasting drought are not the conditions beeches, oaks, and the like crave. They prefer low temperatures even at hotter times of the year. A bit of sun here and there, but apart from that lots of rain and never any temperatures over 77 degrees Fahrenheit (25 degrees Celsius)—that's a dream summer for a tree. While we would give our eye teeth for an accurate weather forecast (at least one that's accurate more than three days in advance), trees turn things completely around. What use is a weather forecast to trees when they can simply make the local weather themselves? They can't do it on their own, though. The larger community of trees in the forest must work together.

I learned all about this in the Holy Halls, where I met with an expert who researches the cooperative behavior in trees.

7

Forest
Air-Conditioning

MY MOST INCISIVE *AHA* MOMENT concerning trees and climate change hit me while we were shooting the documentary *The Hidden Life of Trees*. The film crew and I had an appointment in the Holy Halls with Pierre Ibisch. Ibisch—a likable man who had impressed me when he visited the forest I manage in the Eifel—is a professor at the Eberswalde University for Sustainable Development (HNEE). The Holy Halls in question are not large buildings but one of the oldest beech forests in Germany. A number of the trees growing there are more than three hundred years old, and apart from a few interventions, trees haven't been felled there for about 150 years. The moment you step inside, you're in the embrace of one of central Europe's last primordial forests. Toppled giants are decaying around you, filling the air with the aroma of fungi. In the shadows, multitudes of young deciduous trees are growing infinitely slowly. Once upon a time, all central and western Europe must have looked like this!

Ibisch and I wandered through the preserve with the camera crew, marveling at small wonders everywhere. There was the broken-off giant beech. The only part left standing was a narrow sliver covered in bark. A delicate new crown was growing from this 13-foot (4-meter) -tall toothpick. The skinny remnant was keeping the old tree—and most importantly, the old roots—alive with the sugar it was manufacturing in its leaves.

Then there was the almost completely rotted trunk that looked like a long earthen berm. Even though we'd had no rain for weeks and the fields edging the forest were dry as dust, the fallen giant was moist to the touch. Ibisch invited me to dig my fingers into the crumbling trunk. The rotten wood was like a sponge, and when I squeezed a handful, water ran out. This small ancient beech forest was fairly bursting with moisture, a small miracle considering how dry the previous winter had been.

The *aha* moment in question, however, had been triggered earlier at the entrance to the preserve when Ibisch spread out a couple of maps on a wooden table and we had our first in-depth conversation. The maps he unrolled for me showed various sections of the landscape on the outskirts of Berlin. On one I could see pastures, agricultural fields, forests, and lakes, along with communities, each in a different color as you usually see them on topographical maps. On the other, the same landscape was depicted in all colors of the rainbow.

Ibisch explained that the second map was a temperature map where the temperature gradient was colored as you would expect it to be—from blue to green, yellow, orange, and red. In other words, blue indicated where it was cold and red indicated where it was hot. This map had been created over a period of fifteen years using satellite data. Temperatures were measured in the summer months of June, July, and August, on cloudless days when the satellite had a clear view of the ground. The surface temperature was measured at about noon, when the eye in the sky flew over Berlin, and data was gathered for a total of 470 days. The measurements ended in 2017, before the three record-breaking summers with even higher temperatures. Even so, the results were enough to make your hair stand on end. The map showed that heat waves are caused not only by climate change but also, crucially, when natural forests are destroyed and the landscape is transformed into plantations, arable land, and communities.

The results plotted on the map showed Berlin deep red, while the surrounding lakes were solid blue. This was no surprise. The average temperature in Berlin at noon during the summer months over fifteen years was around 90 degrees Fahrenheit (33 degrees Celsius), whereas the lakes never got above 66 degrees Fahrenheit (19 degrees Celsius). That sounds like a no-brainer. After all, asphalt and concrete warm up much more easily and quickly than large bodies of water. But the difference between urban and country areas is not the most important result from all those measurements. What is much more interesting is what was happening in the surrounding woodlands in the summer. If you glance at the temperature map quickly, it's difficult to tell some of them from the lakes. These cool areas are ancient deciduous forests. The result of the experiment lay right there in front of us: beeches and oaks act like bodies of water. They cool the landscape so much that the temperature difference between a city like Berlin and an ancient woodland is about 27 degrees Fahrenheit (15 degrees Celsius).

Open country with pastures and arable land was up to 18 degrees Fahrenheit (10 degrees Celsius) warmer than the ancient forests. The greatest surprise for me, however, were pine plantations. The results made it clear that these dismal monocultures can never replace real forests. The temperatures in the plantations were up to 14.5 degrees Fahrenheit

(8 degrees Celsius) warmer than those in the ancient deciduous woodlands. And then there is the fact that the crowns of these conifers catch more rainfall and so the ground below is significantly drier.

HAMBACH FOREST SHOWS just how much even a small remnant of forest can benefit the local climate. It is one of the most famous forests in Germany because it has become a symbol of changing energy policies. The forest's fate seemed sealed when excavators digging for lignite in the neighboring open-pit mine chewed their way to within a few yards of the forest's edge. From what was once about 15 square miles (40 square kilometers) of trees, only three-quarters of a square mile (2 square kilometers) remained. The clearing of the last miserable remnant of forest was halted by an injunction from the Münster Higher Administrative Court[1] after the federal and state governments came to an agreement following protests by environmental organizations and activists.

But is it too late to save Hambach Forest? The gaping maw of the enormous, nearly 1,000-foot (300-meter) -deep mine lies at its feet. In summer, hot winds rise from the mine's depths, creating a powerful vortex that sucks away the cool, moist air the ancient trees have put so much effort into creating. Storms sweep over the mine unobstructed, toppling trees at the forest's edge and slowly but surely reducing the size of Hambach. Hambach lies in an agricultural wasteland that heats up in summer almost as much as the open-pit mine. There are hardly any other forests around that could bring their influence to bear and help create a local microclimate more friendly to trees.

So is there any chance the ancient forest can survive? To answer this question, Greenpeace asked a team led by Pierre

Ibisch to undertake a local climate study.[2] You already know the main method they used: measuring the temperature of different land surfaces via satellite and coloring them in on maps. There were other ecological investigations as well. The study found that in the extremely hot summer of 2018 the difference in temperature between the forest and the open-pit mine was up to 36 degrees Fahrenheit (20 degrees Celsius). You cannot help but be seriously impressed that such a small, but reasonably intact, forest exerts such a strong cooling effect.

DESPITE THE RESULTS of the study, the future of the old trees in Hambach is, unfortunately, not particularly rosy. The excavators continue to chew their way closer and closer, and trees around the edge of the forest are dying from the heat as winds from the mine act like a gigantic hair dryer, blowing away the cooling effect of the forest. At the same time, large quantities of moisture are being removed, or, to stay with the same metaphor, Hambach Forest is constantly being blow-dried.

The severity of the situation becomes even clearer when we remember that every day a mature beech releases up to 130 gallons (500 liters) of water into the air via its leaves—water that is hard to find in the ground because of the proximity of the lignite mine. To add insult to injury, the open-pit mine is being emptied of water by giant pumps because the bottom of the mine lies far beneath the water table and, without the pumps, the enormous hole in the ground would flood.

Experts have recommended planting a buffer zone around the ancient forest to save it. At the very least, young trees could reduce temperatures around Hambach Forest a little bit and humidify the air to relieve some of the stress now plaguing the ancient forest. Images taken by Greenpeace show that buffer zones planted around human communities would be a blessing for us too.[3] Members of the environmental group took photographs of Cologne using a thermal-imaging camera. This large city in the Rhine Valley is about an hour's drive from our forest lodge. The results were similar to those from Berlin and Hambach Forest. Buildings and asphalt glowed red in the summer heat. In contrast, the deep blue of trees in city parks made the parks look like lakes, and their temperatures showed they did indeed act like lakes. The green giants reduced the temperatures by up to 36 degrees Fahrenheit (20 degrees Celsius). That's a strong argument for adding more green to cities.

The forest offers us another benefit in addition to its cooling effect. It delivers more rain, as we'll see in the chapter on rivers in the sky. Before we move on, however, there's a glimmer of hope that there might be some movement in the views of state foresters who spend their time felling trees in nature's cooling systems. Ulrike Höfken, then minister of the environment for the German state of Rhineland-Palatinate announced a temporary moratorium on logging in ancient beech forests to the end of 2021.[4]

8

When Rain
Falls in China

FORESTS SHAPE CLIMATE not only locally but also across continents. Water plays an important role. We've already seen how trees cool the air around them. Trees also significantly influence running water.

For starters, trees reduce the amount of moisture that penetrates soil layers to reach groundwater. Some gets trapped in their crowns, and a larger portion is used by the trees to grow biomass and to cool the forest through transpiration. Depending on the species, this can amount to the equivalent of 28 inches of rain (700 liters per square meter) per year.[1] To put this into perspective, the rainfall around Magdeburg, one of the driest regions in Germany, is a scant 20 inches (500 liters per square meter) per year. Forests can survive here only if the trees cut back on their water use by drinking less and therefore growing more slowly than they would in other locations.

Does that mean forests destroy water supplies and dry out landscapes? Not at all, because the water lost through transpiration doesn't disappear. You could say it is recycled and flows into other areas—through the air. Aerial rivers contain water in a far less concentrated form than rivers on land because they are formed from water vapor, but these rivers flow, nonetheless, as shown by studies undertaken by Russian researchers. The scientists were investigating where rain that falls in China comes from.

That might sound like a strange question because, in general terms, rain comes from the nearest ocean. Water vapor rising from these large water bodies forms clouds that are blown across continents by the wind. The clouds release rain over land and water follows gravity down to rivers that flow into the ocean—the cycle is complete. What's important for plant life on land, therefore, is that at least as much moisture comes from the air as is lost when water flows back to the ocean and through transpiration. If this doesn't happen, everything dries out and the region turns into a desert.

The Russian scientists, Anastassia Makarieva and Victor Gorshkov, discovered adequate moisture replacement does not always happen.[2] According to the researchers, the amount of rainfall usually diminishes exponentially the farther you get from the ocean. By the time the clouds have traveled a few hundred miles inland, they have released all their moisture and no more rain falls. And without rain, plant life is impossible. At least this is what happens if there are no forests nearby. Large forests change everything. They suck moist air into the interior of continents—and they do so with such force that the research team describes them as biotic pumps. Even thousands of miles from the ocean, there is no discernible reduction in rainfall over large, naturally growing forests.

The two scientists explain the process as follows. Forests transpire enormous quantities of water from their leaves. For every 10 square feet (1 square meter) of woodland, there are up to 300 square feet (27 square meters) of leaf surface at the level of the crowns, from which the trees exhale moisture from an infinite number of tiny mouths. An ancient beech on a summer day can breathe out up to 130 gallons (500 liters) of moisture,[3] which then cools the forest and escapes into the atmosphere in the form of water vapor. Intense transpiration activity in large areas of forest leads to a mass of rising air, which creates a local area of low pressure. As an area of low pressure has less pressure than the surrounding air, air flows into the area. You could say that forests suck fresh air from the oceans—and they do this over great distances.

This moist ocean air then rises above the forests, cools, and rains the moisture it contains down onto the trees. The total amount of precipitation, according to the scientists, is more than the amount of moisture the trees lose when they exhale. The trees' water use can, therefore, result in them having access to even more water. Forests in Siberia confirm this conclusion. Trees actively transpire water from their crowns only in summer. In winter, when everything is frozen solid and the trees are asleep, the forest water pump should grind to a halt. And the research team observed that this is exactly what happens.[4]

If the forests are cut down and replaced with, say, grass or croplands, rainfall could decrease by up to 90 percent. This sounds plausible, and indeed we can observe this happening today. Since the turn of the last century, droughts have been increasing in the Amazon. This is happening in concert with the disappearance of coastal rain forests, the increase in logging, and the reduction of the size of the rain forest. In other words, if you disrupt the pumps close to the ocean, it should come as no surprise when water no longer reaches the interior. Observations from Germany about the cooling effect of forests and more rainfall over ancient forests support the researchers' thesis.

More compelling evidence supports the idea of forests as water pumps. A team led by Ruud van der Ent from Delft University of Technology in the Netherlands investigated how nature recycles water.[5] The researchers stumbled upon a simple fact: water that evaporates into the atmosphere must, at some point, fall as rain. In the team's view, this self-evident fact is often overlooked in studies by hydrologists. In professional circles, water that evaporates is lost to the system and new rainfall comes from outside. It not only sounds logical that water is passed on in ecosystems on a large scale, it's also extremely important for our understanding of how the green lungs of the planet function. A huge recycling system is in place that functions much better than our society's use of raw materials. As moisture evaporates multiple times to then fall again as rain, it's reused by plants up to ten times over—as long, that is, as forests are not felled over large areas.

When you combine the research of the Russian and Dutch scientists, it becomes clear that forests have a hitherto completely underappreciated importance for water management on our planet. They not only influence wind systems by creating areas of low pressure that draw clouds from the oceans into the interior of continents, but they also continually reseed the air with moisture. When it comes to climate change, many foresters still think of trees as little more than biological storage units for carbon dioxide either while they are alive or, better yet, when they are dead. Every tree that becomes part of a house or a piece of furniture is an environmental success story. The carbon in processed wood can no longer be released into nature by bacteria and fungi, which is what happens when trees die a natural death. The living and breathing entity that is a forest is reduced to a vault full of carbon, its effects on global water and temperature management ignored. If we were to give full credit to trees for the climate services they provide, it would become clear that protecting forests is more important than using wood as a raw material, and that we need to cut back drastically on our consumption of planks and paper.

Water is one of the key factors that make human life possible. In hot, dry regions conflicts flare up regularly around rivers, such as the Nile, that flow through many countries. Egyptians draw 95 percent of their fresh water from this gigantic river, and they would be lost without its water. Additionally, agriculture in the fertile valley would be almost unimaginable without the Nile. Ethiopia, on the upper reaches of the river, has recently constructed a hydroelectric dam. It will take years for the enormous reservoir to fill with water and become fully operational. In the meantime, all this water will be unavailable for Egypt and for Sudan, which shares Egypt's predicament. A war that threatens the area has so far been avoided through international mediation.[6]

When people realize how important the aerial rivers managed by forests are, it's conceivable that one day there could be conflicts here too. There is just one snag. You can open dams to restore access to more water to the people who live downstream. When it comes to forests, it's not so easy to reactivate aerial rivers after forests are lost to clear-cuts. Even if you were to replant the deforested area, it would take decades before the new forest began to regain its former function. We can observe a big project of this kind going on in Brazil right now. The reforestation of coastal rain forests has begun, albeit in a patchwork fashion. It remains to be seen how long the process of regeneration takes—even in a tropical region where trees grow exceptionally quickly—and whether the pump ever really begins working again as it should.

I wish this time around people would pay more attention. This is the second time the forests' effects on temperature and water cycles have been discovered. The famous explorer and scientist Alexander von Humboldt described the importance of these connections in detail way back in 1831. He wrote in *Fragments of a Geology and Climatology of Asia*: "The rarity or lack of forests increases both the temperature and aridity of the air, and this aridity, by lowering the rate of transpiration and the vigor of the grasses, has a deleterious effect on the local climate."[7]

IS IT JUST A COINCIDENCE that trees work together to both cool the air and make their own rain? Forests having been growing for the past 300 million years. We know the extent to which the green giants work cooperatively, warn each other of danger, and share food and even memories via their roots. Therefore, I believe this huge community of large plants has managed to move beyond passivity to take at least some aspects of the weather into their own hands—or should I say, leaves. The fact that right now so many trees are dying in hot summers doesn't contradict this. Quite the opposite, in fact. This forest die-off shows what happens when we disturb this perfectly tuned community. Foresters hack forests to pieces, create clearings, and change the character of the forests by introducing unsuitable species of trees. What are now, from a global perspective, mere fragments of primeval forests no longer function properly. As we continue our forest walk, I will tell you how we can reverse this process. Yes, it can be done!

If communities of large plants work cooperatively to influence the local climate, it makes sense that they probably look after each other in other ways. There are intriguing new scientific developments on this subject that I am excited to share with you now.

9

Take Care and Stand Back

THE TERM *MOTHER TREE* comes from forestry. It has been clear for centuries that tree parents play such an important role in raising their offspring that they can be compared to human parents. You might remember from my book *The Hidden Life of Trees* that a mother tree identifies which neighboring seedlings are hers using her roots. She then, via delicate connections, supports the seedlings with a solution of sugar, a process similar to a human mother nursing her child. Shade provided by parents is another form of care, as it curbs the growth of youngsters living under their crowns. Without the shade and exposed to full sunlight, the young trees would shoot up and expand the width of their trunks so quickly they'd be exhausted after just a century or two. If, however, the young trees stand strong in the shadows for decades—or even centuries—they can live to a great age. Shade means less sunlight and therefore considerably less sugar. The slow pace of life gently imposed by the mother tree is no accident, as generations of foresters have observed. To this day, they talk of what is known in German as *erzieherischer Schatten* or "instructive shade"—in other words, shade cast intentionally by a mother tree.

Mature trees also help each other later in life by pumping sugar solution back and forth through their roots so that weak and sick individuals can survive hard times and later regain their health. Then they once again contribute to cooling the forest, which benefits all the trees equally. In times of climate change, it is therefore especially important to leave these forest communities undisturbed. The same goes for trees that look as though they are dying (and most often are only sick).

The assistance trees extend to each other likely goes beyond the processes I've just mentioned. Students at Aachen University conducting studies in the forest I manage have discovered there are barely any differences in the performance of individual trees in an undisturbed beech forest. For instance, all the trees seem to photosynthesize at about the same rate, with no particularly weak ones and no particularly strong ones. In contrast, in ancient beech forests that are exploited commercially, where many trees are felled, the remaining trees seem to become selfish. In these forests, some trees are strong and others are weak, and their ability to photosynthesize varies widely. The differences in performance are hardly surprising, given that there are literally no longer any points of contact between them—not through their roots or their leaves. Selfish trees don't help each other, maybe because the gaps between them are simply too large. Perhaps they are not being selfish after all but have become loners forced to survive without any direct help from their neighbors.

EXPERIMENTS USING ARABIDOPSIS (*Arabidopsis thaliana*), a common laboratory plant whose genes have been thoroughly researched, show how consideration for others could have arisen in the plant world. Arabidopsis is easy to grow in petri dishes, reproduces quickly, and produces a multitude of seeds. Topping out at about 1 foot (30 centimeters), it also remains fairly small. Trees, which grow to be 100 feet (30 meters) or more, just can't compete in the height department. You could say arabidopsis is the lab rat of the plant world.[1]

María A. Crepy and Jorge J. Casal in Buenos Aires, Argentina, are two researchers who grow this plant in their laboratory. They have established that the plants show consideration for each other when they position their leaves. When plants grow close together, their leaves shade those of their neighbors. This reduces the neighbors' ability to photosynthesize, and they end up with less to eat. Naturally, this weakens the neighboring plants, and in terms of competition that would have to be an advantage—after all, plants usually battle for light. But the research results clearly demonstrated that the plants were not always competitive. When arabidopsis recognized relatives, its behavior changed. As soon as a plant registered that its neighbor was a member of its own family, it considerately adjusted its tiny leaves so its neighbor did not have to starve in its shade.

Does that sound completely off the wall? Actually, it would make much less sense if being considerate to a family member were a purely human characteristic. Recognizing family relationships and acting considerately to relatives are common phenomena in the natural world and make perfect sense. Everywhere the survival of individuals depends on the strength of the community, organisms work in teams. Mammals live in family groups and herds or flocks, birds pair up for life—ravens are a good example—and even some unicellular slime molds collaborate to form fruiting structures.

But how does arabidopsis recognize a family member? If we think of trees and their social networks, the answer is obvious: it could well be through the roots. After all, we've known since the 1990s—thanks to research pioneered by Canadian scientist Suzanne Simard—that the giants use their roots to share food, exchange information, and even recognize their own seedlings. But Crepy and Casal decided to make things difficult for arabidopsis. Every tiny plant was given its own pot and isolated from its neighbors. Then the pots were placed so close together that the leaves from one plant shaded the leaves of others. And this is where things got interesting. When the plants were related, they directed their leaves away from each other. The research team discovered that arabidopsis recognized relatives using special wavelengths of red and blue light. In other words, they recognized relatives by sight. The two scientists tested to see if light waves really were involved by designing an experiment using specially bred plants that did not have the receptor for these wavelengths. And voilà: these plants ignored relatives, clearly because they could not see them.

Arabidopsis is not particularly speedy. It takes a plant many days to move its leaves to one side to help out its neighbors. Once it's done this, the neighbors have more light. But what advantage does this sort of considerate behavior have for the thoughtful plant itself? After all, its leaves were in the optimal position. After adjusting them to benefit of their relatives, it was now shading of some of its own leaves. However, because the neighbors were showing the same consideration, the total amount of light falling on the plant's lower leaves increased. More light equals more energy, which in turn equals a higher level of fitness. In this study, arabidopsis plants growing with family members produced more seeds and therefore were more successful.[2]

WHEN IT COMES TO FAMILY, are trees as considerate with their leaves as arabidopsis? This issue has not yet been resolved, but there is a phenomenon that has been raising questions along these lines for the past century: so-called crown shyness. If you're in a deciduous wood on a summer day and look up into the trees' crowns, you'll sometimes spy a narrow gap around the branches of individual trees—often less than 20 inches (50 centimeters) wide. It looks as though no tree wants to fill this border zone with its own leaves and branches. From the air, whole forests often look as though a delicate network of care and consideration spans the spaces between the trees.

But is this really the trees respecting each other's space or just the effect of the wind, as many researchers suggest? Their hypothesis is that the outermost branches rub against those of the neighbor as the crowns sway until, eventually, they no longer intrude into each other's territory.[3] According to these scientists, it has nothing to do with one tree respecting another tree's personal space; it is a purely mechanical phenomenon. My own observations do not support this argument—and you can check my observations out for yourself every time you take a walk. You can see tree branches intruding into other trees' space all over the place. Limbs touch and some trees even grow branches into their neighbors' crowns. Wind and storms blow everywhere, so the results of branches rubbing up against each other (which undoubtedly is responsible for some branch loss) should be clearly visible in every forest. And yet it isn't—you have to go looking for crown shyness.

If crown shyness really is an example of the same consideration for others demonstrated by arabidopsis, there's something that might explain why trees don't practice this considerate behavior all the time: most of our forests are planted. The seed suppliers mix the seeds thoroughly before they send them out to tree nurseries. Obviously, when the little trees are planted out in the forest, they end up next to complete strangers. Only in places where natural forests grow—for example, beech families in large groups that have been growing together for centuries—would you expect to stumble upon crown shyness more often. I don't know of any research to support this idea, but I will keep an eye out for this behavior when I visit ancient forests in the future.

A TEAM LED by biologist Roza D. Bilas concluded in a review article that recent data contradict the idea that plants are mostly passive actors in their environments. The team also states that it's unlikely plants have spread around the world for the past 500 million years without being able to recognize and react to other plants, whether they be friends, neighbors, or foes.[4]

Trees live in relationships—and not only with their own kind. The very smallest of life-forms are an important component of the forest community, even if we are only just beginning to take notice of them. And now we, at least, dear readers, will change that.

10

Underrated
All-Rounders

IT'S FUN TO ENGAGE CRITICS in a dialogue, and that's the reason my son, Tobias (head of our forest academy), and I invited one of my greatest critics to Wershofen. It didn't take long for us to get into a heated discussion that finally led to the question of species diversity in the forest. The university professor and forest scientist, who agreed to meet only if no media were present, is a fervent defender of forestry. In his opinion, nature benefits when foresters create clearings by felling trees. He made the sweeping statement that when the sun streams in and warms the remaining trees, species diversity increases significantly.

I can't help but smile when I hear such assertions, and I'm not the only one who finds them fundamentally unscientific. To determine whether species diversity has increased, you must first identify precisely what species are there—which means you have to count them. After trees have been felled, you can count them again and thus determine by a simple mathematical calculation whether the total species count is higher or lower than before. The problem with this is, we haven't the slightest idea how many different creatures are milling around in our local ecosystems.

A team led by Kelly Ramirez from Colorado State University gives us some idea of the multitudes that are bustling about in the ground alone. The researchers took about six hundred soil samples in New York's Central Park and analyzed the genetic material they contained. The team found traces of 167,169 different species—all microorganisms about the size of bacteria, of which about 150,000 were hitherto unknown.[1]

When I meet researchers, I like to ask them how many species they think we have yet to discover. The results of my informal survey are about 85 percent. That means we know an estimated 15 percent of all species in Germany. The results are likely much the same around the globe.

Back to my conversation with the forest scientist. I asked him if he agreed with the other scientists about the estimated number of as-yet-undiscovered species. "Oh, you mean bacteria and fungi!" he replied dismissively. Clearly, he felt those life-forms and the research being done on them were barely worth mentioning. But those who don't know about bacteria and the like cannot fully assess the effects of interventions in ecosystems, especially not when it comes to the decrease or increase in species diversity. A team led by US researcher Roland Rodriguez has stated: "Our limited understanding of such important microorganisms is a testament to the fact that the 'age of discovery' is just beginning."[2]

And the little guys are important! The human body is a good example. We have at least as many bacteria bustling about inside us as we have body cells. These bacteria are as much a part of us as our blood cells or nerve cells. Recent research has shown how much influence they have—gut bacteria, for example, produce substances that communicate with our brains. In short, bacteria have a big say in our lives. They can affect our behavior by making us fearful or depressed.[3] Thomas Bosch, leader of a research team at Kiel University, goes even further. He suggests that the human nervous system arose not to tell parts of our bodies what to do, but so our bodies could communicate with microbes.[4] And so the expression "Listen to your gut" is suddenly elevated from adage to hard science.

Every one of us is a small ecosystem with a unique combination of thousands of species of bacteria, as individual as a fingerprint. On the surfaces of our hands alone, each of us hosts an average of 150 different species. Your left hand is so different from your right hand that only 17 percent of the species on each are identical. The different species of bacteria on the hands of different people match only 13 percent of the time. In total, researchers found 4,742 different species on the surfaces of their subjects' hands. Let's just quickly compare that with species diversity in vertebrates: in all of Europe there are about five hundred species of birds.[5] The skin on your hands, therefore, is a hot spot of biodiversity. Incidentally, washing your hands doesn't disrupt this little cosmos. The tiny beings reproduce so quickly that it doesn't take them long to go back to the way they were.[6]

As we cannot survive without these microorganisms and they cannot survive without us, science has come up with a new classification and now refers to us as holobionts (holo meaning "whole" and bios meaning "life"). When you think of Earth as being populated by holobionts, it sounds as though we are stepping into a science-fiction film. And yet, in many cases, our previous clear demarcation between separate individuals no longer makes sense—at least not for multicellular species. And with 100 trillion cells in our bodies, we humans are indeed a multicellular species.[7] This means the term *species diversity* itself is not very useful, as within each species there's a multitude of holobionts—every individual is different.

The idea that a physical body comprises a unique ecosystem containing thousands of species is probably true for all multicellular organisms—and it is certainly true for trees. That should—no, *must*—cause us to radically change how we view forests and how we interact with them.

Pierre Ibisch from the Eberswalde University for Sustainable Development clearly states this new way of thinking: "Ultimately, it appears that it is not biological species that are the subjects of ecological interaction and evolution, but rather complex holobionts. We are on the threshold of a completely new understanding of forest ecosystems and the entire living world. Incredibly large blind spots are emerging. And this at a time when people are delving more deeply into the fabric of ecosystems, disrupting them more thoroughly and on many more levels than they have ever done before."[8]

WHENEVER YOU BEGIN to lose sight of the big picture, it's important to step back for a moment and reflect on what we know. Each time biologists discover something new, our view of what is happening in the natural world becomes slightly more opaque. To put it more precisely: modern research reveals that we've never had a clear picture of what is happening out in nature.

Making fine distinctions between categories and allocating functions to species within ecosystems is difficult to do in the real world and is problematic anyway. This kind of mapping arose from an idea of nature from centuries past, which viewed our environment as a carefully calibrated machine. Every species has a function assigned to it at birth that it fulfills for the rest of its life. Whatever its function is, the species is usually thought of in terms of how useful it is—usually, how useful it is for us. Whether organisms are categorized as beneficial or harmful always depends on how much they promote or impede human interests. And that's the crux of the matter. This way of seeing the world puts humans at the center. Only humans have no specifically assigned functions. All other life-forms are cogs in a machine that has us as the crowning glory of creation.

To understand how the machine works, scientists divide it into different cogs, that is to say, different species. But nature cannot be deciphered this easily once the term *species* has been toppled and we realize it is all about holobionts—that is to say, ever-changing ecosystems—such as the ecosystem each one of us comprises. Bacteria, which triggered this whole upheaval, are now in the hot seat. Can the different species of bacteria ever really be designated as species?

According to the old definition, to constitute a species, life-forms must be able to reproduce sexually and produce fertile offspring. But bacteria don't do this. They simply divide without much fanfare, thereby raising the question of whether they are now two new bacteria or a mother and her offspring. Additionally, the two are often genetically very different. Whereas most scientists agree that human DNA differs from that of chimpanzees by a mere 1.5 percent, bacteria that supposedly belong to the same species can differ genetically by as much as 30 percent.[9] Why does science allow this for bacteria and, quite rightly, not for animals? Because if it didn't, we'd never be able to divide bacteria into species. This example shows that science can no longer control the enormous diversity of living beings.

And if that was not enough, bacteria, in turn, are colonized and eaten by viruses. Researchers calculate that once tucked inside their prey, about 30 billion bacterial viruses a day(!) get a free ride through our gut membrane and into our bloodstream, and from there into every conceivable organ.[10]

Ew! Have you lost sight of the big picture? I have, but truth be told, it really doesn't matter. Simply admitting that we are nowhere near understanding the cycle of life is both freeing and humbling—especially the latter for those of us who have tried to redesign nature to function better with rather than without our intervention. The take-home message is simple. If we want to continue to enjoy nature in all its glorious complexity, we must simply observe. Locally extirpated species of plants and animals can be reintroduced here and there, of course, but after that—hard as it is for people who care deeply—basically the only way to restore an entire ecosystem is if, after a small initial boost, we simply leave the area alone.

BUT I DIGRESS. The cooperation between bacteria and plants—in this case, specifically trees—and the way they merge to form a single organism are nothing new. Do you remember biology lessons in school? Students learned (and still learn) about bacteria in root nodules. These and a few other bacteria perform an important task for plants: they transform atmospheric nitrogen into fertilizer. Humans with their chemical factories are the only other life-forms that do this. Without bacteria, trees would have to depend on lightning and volcanic eruptions—and naturally occurring wildfires. In these three processes, heat transforms atmospheric nitrogen into a form plants can use. As none of these processes occurs very often, a few species of bacteria stepped up to help trees in their time of need. These nitrogen-fixing bacteria are not acting altruistically, because they can't feed themselves without the trees' help. The tiny organisms need partners that will pay them for their services with food. The word that immediately comes to mind here is *symbiosis*, which means cooperation between two different species.

The relationship between ants and aphids is an example of symbiosis. The ants stroke the aphids' bodies with their antennae, which causes the aphids to excrete a delicious sugar solution. In return, the ants protect their little green flock from hungry ladybugs. Despite their mutually beneficial relationship, both parties—aphids and ants—can survive independently.

Relationships like the one in which fungi and algae combine to form lichen used to be called symbiotic. However, a lichen exists only when both partners combine, and once they have done so, neither can survive without the other. We no longer characterize this relationship as symbiosis, and lichens are increasingly regarded as holobionts. Under the old definition, we could also claim that the scavenger cells in our bloodstream that attack and destroy pathogens are not actually part of us.

The bacteria in root nodules can live independently, at least initially, before they throw their lot in with trees. To attract these little helpers, the tree roots exude nutrients into the surrounding soil as a kind of bait. The bacteria then move toward the finest part of the roots, the root hairs. This is where it gets interesting. When the root hairs and bacteria recognize each other, the tree allows the bacteria inside. This, for me, is the point at which symbiosis ends and the different life-forms combine to create a new entity (a holobiont). The tree now builds the newcomers a comfy place to live by forming nodules on its roots. This costs the tree energy, but the bacteria will pay it back by providing the tree with nitrogen fertilizer.

Trees with root-nodule bacteria can colonize soils that are naturally lacking in nitrogen. And because trees grow taller than grasses or forbs, merging with root-nodule bacteria offers them the benefit of a literal leg up. Many species of alder take advantage of this benefit, as do locust trees. Many trees, however, are unable to cooperate with these bacteria, and others could but don't. The European hornbeam is one example of a tree that so far has not allowed the tiny life-forms to enter. We do not yet know the reason for their reluctance.[11]

Cooperative behavior between trees and bacteria also takes place outside the roots. To date, there has been no detailed research on what is happening, but what has been discovered is tantalizing. According to researchers at the Netherlands Institute of Ecology in Wageningen, plants, like us, have an immune system so they can defend themselves from pathogens. Unlike us and other animals, at least some of this immune system is located not inside but outside their bodies. This immune system comprises a community of bacteria around the tree's roots that prevent the roots from being infected with soil-borne pathogens such as fungal root rots.[12]

LET'S RETURN TO our conversation with the forest scientist who came to visit us in the Eifel. He clearly attached no importance to these complex relationships between life-forms because during our conversation he described an ecosystem as being of good quality based solely on the quantity of known species. But if we don't know about an estimated 85 percent (or even considerably more than that), then we can't count them, which means we can't use quantity as a criterion to evaluate ecosystems. When we don't know most of the species, there is no way to make a scientifically sound argument that biodiversity increases when humans intervene. And yet the easily refutable, feel-good tales of a benevolent forest industry that gives biodiversity a boost through clear-cuts and plantations are still widely taught in universities. But luckily there's a remedy is in sight, and I'll come back to it later.

Scientists who decline to expand their knowledge base are nothing new. In the case of forest science, this is particularly tragic. Trees are key to slowing down climate change, and right now forestry is having a negative impact on two-thirds of the world's forests.[13]

When you think of the complex communities that keep forest ecosystems going—with their mind-boggling number of different, mostly small life-forms—you can see that the forest industry is acting with all the finesse of a bull in a china shop. Its answer to climate change is to then replace what it has broken. Species are swapped out. For instance, beech forests are replaced by plantations of nonnative edible chestnuts or cedars of Lebanon. With this kind of intervention, forests become artificial constructs far removed from nature, and the risk increases that they will be far less capable of withstanding climate change. Why precisely those people who are supposed to be protecting and preserving our forests have taken such a wrong turn is an issue we will turn to in the next chapter.

PART II

When Forestry Fails

11

Backed Up
Against a Wall

FORESTRY AS IT has been traditionally practiced is facing massive problems. Spruce and pine plantations are dying, and the public is beginning to notice this is not all due to climate change. Bark beetles are eating up the monocultures of trees, fire is laying waste to forests, and chain saws are severely weakening the trees' amazing ability to generate rain and cool themselves.

Everything was once so great and worked so well for such a long time. Many countries around the world followed the example of foresters in Germany and transformed large sections of their forests into plantations, a practice that provided a reliable source of wood for the timber industry for decades. Switching to fast-growing species and breeding trees for desired traits brought results like those achieved by factory farming: individuals ready for harvest at a young age, all with a relatively uniform carcass weight.

Just like animals on factory farms, however, trees in plantations are extremely delicate, and substantial losses due to diseases and natural events are a constant problem. And wood from these tree factory farms is of significantly lower quality than wood from primeval forests. The public is not aware of this because the industry has adjusted to thinner trunks and lower-quality lumber. Technology has stepped in to replace what trees can no longer offer because of poor forest management practices. Just try to buy a thick wooden beam made in one piece. It's practically impossible. Nowadays, thick beams are made by gluing tiny boards together so you can produce beams of any size without the need for large trunks.

Everyone seems happy and no one notices that brutal management practices are making the forest increasingly fragile. Climate change is the straw that has broken the camel's back, starkly revealing the problems of the last decades. The beautiful house of cards constructed by state-organized and -planned forestry is slowly and inexorably collapsing.

Despite the similarities, it is much more difficult to plan in forestry than it is to plan in agriculture. The products share many similarities. Wood is a perishable commodity. After it is harvested in summer, it often must be processed within a few weeks before wood-destroying fungi or insects significantly reduce its quality. The pressure doesn't let up even in winter, as in these times of climate change this season is becoming so warm that fungi can grow in wood then too.

One big difference between forestry and farming is the length of time between planting or seeding and harvest. Whereas farmers can change their plans every year, once foresters have decided what to grow, depending on the species, they are tied to their decisions for the next sixty to two hundred years. But who can know so far in advance what the market will want? And then along comes climate change, which greatly increases uncertainty. Now it's more than simply a question of what the future demand will be; it's also a question of whether the trees will grow large enough and reach a suitable harvest age before they die. And as if that weren't enough, even without climate change, here in Germany winter storms blow through every few years, flattening large numbers of trees. And because wood is perishable, these trees must get to market quickly, which means prices plummet.

On top of all that, of course, comes the question of sustainability at this scale. A farmer simply begins again the year after a catastrophe. However, after a storm blows through, forest owners must hold back from logging the remaining trees since, as far as German law is concerned, the storm has felled way too many trees prematurely. Then drought years with their attendant attacks by bark beetles are a regular occurrence, as are fashions trends that suddenly favor different types of wood for furniture. In the worst-case scenario, entire markets collapse, as happened when wood was no longer needed to brace tunnels in mines.

To sum up: in forestry few predictions can be made about what will happen in the long term. Despite this, under German law, the owners of larger private forests and managers of public forests are required to draw up ten-year plans. Calculations are made, plans are formulated, and measurements are taken only to find after ten years that once again everything has changed. I have yet to see a single case where these long-term calculations made any sense at all.

There's another, completely different reason long-term planning in forestry gets derailed. Even forests that are still reasonably intact are producing less wood. The reason for this decline makes perfect sense, even to those who know nothing about forestry. Trees that discard their leaves prematurely in summer cannot grow as much wood as they would in a normal year. And if the situation overall deteriorates to such an extent that we hardly ever have normal years, then plans should be modified to reflect this—or, rather, they might have to be modified.

During discussions at the forest academy, we have seen over and over again that foresters react as though climate change is an accounting problem. Accordingly, they strike the dying stands of spruce from their books. Their new plans glide over the fact that the remaining beech and oak forests are also struggling—and they immediately focus on them. This weakens the forests best placed to withstand climate change: ancient stands of oaks and beeches. The trees' social connections are destroyed; the forest floor heats up in the sunshine and dries out. Unfortunately, some of the most impressive trees now die. The forest agencies, however, are never at a loss for words where good public relations are concerned. Beeches? Dying? Then we'll just blow them up—at least that will make for good headlines!

12

Butchery in the Beech Forest

IT'S A SUNDAY in September 2019 in the Thuringian Forest and explosions are echoing down the valleys. Old beeches groan as they lean to one side before hitting the ground with a resounding crash, their crowns splintering into a thousand pieces. The army is at work. Soldiers are laying explosive charges at the bases of the ancient giants, then blowing them up.[1]

That day, thirty beeches and two spruce trees were felled in spectacular fashion. Media headlines conveyed the clear message that the authorities were responding to the forest crisis, albeit in a somewhat heavy-handed fashion. To set the charges, the explosives experts had to fuss with detonators around the trunks. If the trees really had been so rotten that they could have fallen over at any moment, no one would have dared go anywhere near them. If it was safe to work close to the trunks, you wonder why they didn't simply attach steel cables to a winch and pull the trees down with a tractor from a safe distance. I have a sneaking suspicion using explosives was a way to draw attention to the fact that the authorities were springing into action.

You hear something similar—minus the military—up and down the country. Ancient beeches are falling ill, and people are rushing to take them down. A looming threat is being removed—or so they say. Branches could fall or whole trees might tip over. People could be injured. The largest and heaviest timber-harvesting machine the world, which goes by the spine-chilling name Raptor, is being deployed. The 70-metric-ton machine uses its articulated arm to effortlessly saw down ancient trees whole and lift them to logging roads, where it chops them into little pieces. Raptor cuts down up to eighty old and ailing beeches a day as eats its way through the ancient forests.

But not every tree that looks weak will die. Sick trees can effect a complete recovery. Even when whole chunks of their crowns die, many trees grow replacement crowns a little farther down their trunks. These trees can continue to grow old for centuries. Many individuals that stand leafless in August are completely capable of leafing out normally the following spring. They learn—as we now know.

IN MANY PLACES these trees, which are fighting for their lives and learning to adapt, are deemed to be hazardous and are removed even from deep within the forest. People cite the legal duty of forest owners to protect visitors from danger. But this duty doesn't apply here, as determined by a federal court on October 2, 2012.[2] Ailing trees, the court said, may remain even along forest paths. Forest owners are liable only if they cause the dangers themselves—for instance, by having an unstable pile of logs or leaving a downed tree over a path where it might upend a cyclist. So I think the point here is not to protect walkers. It is simply an excuse to keep cutting down trees for their wood even in forests that are struggling to survive.

Another reason for frantically felling trees in dying forests is without doubt rooted in emotions. When plantations that have been carefully thinned for years or even decades suddenly sicken, it's a highly visible signal that forest management practices have failed. When large quantities of dead trees are left standing in forests, the public is prompted to ask whether we still need foresters and, most importantly, who is to blame for this miserable state of affairs.

In the case of the dying spruce and pines, the authorities and forest scientists in charge refuse to accept any blame. The destruction wrought in cities by the Second World War, they say, led to increased conifer plantations in the postwar years. Germany needed to rebuild. How can you fault those who wanted to get the timber industry back up and running as quickly as possible? This argument is easily dismissed. If people needed construction lumber in the 1940s and 1950s, they weren't going to get it from freshly planted, knee-high spruce trees. And just a few years ago, leading figures in forestry were issuing dire warnings about switching too abruptly from coniferous forests to deciduous ones. Professor Hermann Spellmann, for example, commented as recently as 2015 that planting fewer spruce and pines when reforesting areas was courting disaster. He called for a stronger focus on conifers. And here's something I found interesting about Spellmann: until 2020, he was chairman of the Scientific Advisory Board for Forest Policy at the federal Ministry of Food and Agriculture. His word, therefore, carried a great deal of weight when deciding what forests of the future would look like.[3]

It seems hardly anyone connected with forestry wants to admit they have made mistakes, and now their behavior is being supported by dying beeches of all things. As you know by now, these majestic deciduous trees are under most stress in places where intense logging has destroyed their social networks. The remaining old warriors are fighting for their lives. When beeches die in the summer heat that now fills thinned forests, foresters claim they are not to blame for the poor condition of our forests. Central and western Europe is the land of primeval beech forests. If even native species give up, the problem certainly can't lie with forestry—hurray!

There's no shortage of proffered solutions. The official line is that the trees are failing. Therefore, instead of replacing personnel, let's replace entire forests. That sounds completely mad. And yet this is already happening over vast areas. This rolling up of sleeves, this "We've got this!" attitude, is a great way for the politicians in charge to showcase their wisdom and competence—in a world where trees just want to be left alone.

13

Germany's Search for the Supertree

IT'S MARCH 2019. Julia Klöckner, Germany's federal minister of food and agriculture, is standing in a clear-cut in the idyllic district of Havelland. She's using the planting tool in her hands to tuck one Douglas fir after another into the ground. That, suggested media photos printed after the event, showing the minister holding the North American seedlings, demonstrates both drive and determination.[1] In fact, her actions were emblematic of something completely different: of a business-as-usual approach, of a stubborn ignorance of the fact that the time for establishing new conifer plantations is long past.

There's an oft-repeated saying: "Insanity is doing the same thing over and over again and expecting different results." Another definition of madness could be "traditional forestry." In German forestry today, there's hardly any discussion about changing forest practices, only discussion about changing the forest to fit the practices. What we have right now is a kind of casting call for trees billed as "Germany's search for the supertree."

But can you rejuvenate forests simply by swapping out the trees? Surely not. If you did that, all species in the forest would starve, as we can see if we take a look at human nutrition.

GRASSES ARE A MAJOR SOURCE of the food we eat. Our main source of nutrition is grass? The statement seems incredible but is easily explained. Sweet corn, wheat, oats, barley, and rice all belong to the sweetgrass family. This is not a complete list of edible grasses, but it shows that they play a vital role in our everyday lives. Globally, cereals make up over 50 percent of our diets.[2] Add to that their use in animal feed and you can see that grass seeds also end up on our plates transformed into eggs, dairy products, and meat.

Imagine if the federal government attempted, in the coming years, to switch our food source from tried-and-true cereals to species such as ryegrass, meadow fescue, or soft meadow grass. Our food system would break down because these grasses are completely unsuitable as sources of human nutrition. Consequently, if these (fictional) plans were acted on, we would starve. Any government that toyed with the needs of its constituents like that would be voted out at the earliest opportunity.

Grasses and trees have one thing in common: both are very broad scientific categories. If you draw conclusions based on either of them, you are oversimplifying things. What seems obvious in the case of grasses is often completely overlooked in the case of trees. Trees also serve as dietary staples for thousands of species of animals, fungi, and bacteria—whether through their blossoms, fruit, leaves, bark, and wood, or the humus these turn into. If, for example, you replace native oaks or beeches with nonnative Douglas firs, red oaks, or edible chestnuts, you are condemning hordes of soil-dwelling organisms to death by starvation. Many of them simply cannot digest this exotic new vegetation.

Trees are the first link in the food chain in the forest, and this chain has become highly specialized over thousands of years. Unfortunately, the forest food chain is not always obvious. We are used to the animal kingdom, where food chains are typically organized from small to large. The largest animals, such as large herbivores or especially large predators, are usually at the top. That holds for ecosystems such as oceans and savannahs. If the final links in the food chain are still there, then the ecosystem must still be intact. After all, those at the top of the food chain survive only when all the preceding links are also still present. All you need to do is check out the largest animals, and you get a rough idea of how nature is faring.

IN THE FOREST, however, it's the other way round. Here, the largest life-forms are the first link, and therefore it's easy to overlook the numerous links that form the rest of the chain. The misunderstanding is so deeply ingrained that many people (including many experts) believe that a forest is mainly a collection of trees. This is reflected in laws that define forests as areas covered in trees. If Douglas firs, edible chestnuts, spruce, or pines are growing, then it must be a real forest, even if it is nothing more than a green desert for thousands of native species.

When you follow this line of reasoning, it makes perfect sense that you can simply plant a forest. All you need is a sufficient number of tree seedlings, and if the species you have been planting no longer work for your forestry plan, you can simply choose different ones. This is the ultimate admission that forestry functions like agriculture—every once in a while, you switch out "the crops." The only thing that is different is that the production cycles are longer and therefore riskier.

The new tree species need one characteristic above all: they must be able to withstand current changes in heat and drought. Foresters look for growing zones already experiencing the temperatures and rainfall predicted for this country in the coming decades, setting their sights on trees growing a couple of degrees of latitude farther south. When goals are simplified like this, the choices are easy. Apart from Douglas firs from North America and edible chestnuts from the Mediterranean, Turkish hazel (from southeast Europe and southwest Asia) and Oriental beech (which grows from the Balkans to Iran) are hot favorites. Along with other exotic species, they are expected to satisfy the predicted demand for wood in Germany for the next eighty years.

It makes you wonder why there is still talk of planting large areas with conifers and why forest scientists and the authorities responsible for forests continue to deny any responsibility for the death of these plantations.

Even if deciduous trees are planted, however, that's not a win for native ecosystems if the species are imported. Some are really impressive trees. Take the paulownias, which are also sometimes called empress trees or foxglove trees. Paulownias are hardy to temperatures as low as minus 4 to minus 40 degrees Fahrenheit (minus 20 to minus 40 degrees Celsius), grow up to 13 feet (4 meters) taller each year, and can grow up to 17.5 cubic feet (0.5 cubic meters) of wood in ten years. For comparison, the average tree in Germany is seventy-eight years old and takes a whole lifetime to grow the same amount of wood. The paulownia is, therefore, turbocharged—and it's really pretty too.

DESPITE ALL THE frantic efforts and apparent solutions for our future forests, forest agencies cannot gloss over the fact that all of this has much more to do with supplying future demands for raw materials than it has to do with ecology. It's gradually filtering down to nonexperts that forest restructuring is not only figuratively but also literally much like restructuring a factory—in this case a timber factory. What looks to us after a few years like a perfectly presentable young forest is a complete catastrophe for the forest ecosystem. For many native animal and plant species, the planting of unfamiliar species results in the wholesale removal of everything they rely on to survive. The new trees are empty shells that convey the visual impression of a forest, while the living, breathing content—thousands of native species—is mostly absent. Only a few generalists will survive, species that are not at risk because they are at home pretty much anywhere.

After all is said and done, forestry remains trapped in a traditional plantation system with a limited selection of species. In contrast to earlier decades, however, a well-informed and increasingly critical public is now watching forest restructuring. While this pressure is not yet changing the system, it is changing the language that is being used. There's a lot of creative thinking and word games in forestry these days. Wouldn't trees that grow farther south eventually migrate north anyway? Isn't planting these heat-loving trees giving the forests a helping hand? *Assisted migration* is the tree-friendly term in the new lexicon of officialdom. Translated, this means that we are simply helping those species that sooner or later would have made their way here on their own. The trees are just moving a little too slowly for the pace set by climate change and need a bit of help.[3] That sounds reasonable, so let's look at these measures through a couple of different lenses.

THE FIRST LENS through which to view the introduction of exotic species is the lens of climate change. When climate zones shift, plant life shifts as well. This was what happened in the last ice age. When the glaciers receded, a tundra landscape arose covered with grasses, lichens, and bushes. This was replaced by forests of spruce and pine. As temperatures continued to warm, the conifers were edged out by oaks and, finally, by beeches. This migration of trees following the ever-diminishing glaciers continues to this day. Beeches, for instance, have traveled as far as southern Sweden. And spruce forests, in the vanguard of the trees' march northward, have already made it as far as Lapland. Already? Trees do indeed travel slowly, advancing generation by generation, taking thousands of years to cover hundreds of miles.

In times of climate change, however, completely different criteria apply. Right now, it takes less than a decade for a climate zone to shift. The only trees capable of keeping up are those with seeds that drift long distances. If they get caught in a decent summer storm, the tiny seeds of poplar and willow, packed in cottony fluff, can cover more than 60 miles (100 kilometers) in just a few hours. Beeches and oaks are at a disadvantage, as their heavy seeds fall directly at the feet of the mother trees, no matter how hard the wind is blowing. A few birds—jays, for example—can move the trees' seeds a few miles farther (where the birds bury them as winter provisions). The average speed at which these heavy-fruited species travel is about 1,300 feet (400 meters) a year. That used to be fast enough to change locations as the climate changed. Today, however, they are definitely far too slow.

And there is another, much more difficult obstacle all trees must overcome: human ownership of property. If trees want to migrate north, they need to be allowed to grow in pastures, agricultural land, and cities so they can gradually shift the boundaries of their home territories. But who would be happy for their lawn to be temporarily taken over—temporarily in this case would be for a century or more—by trees making their way north?

Any tree that takes root where it is unwelcome is removed as quickly as possible. I completely understand. Many large trees grow around our forest lodge. We also have grassy areas where we like to sit and drink coffee or play badminton. To allow the land around our lodge to be completely overgrown with trees would be a step too far, even for me. And because we all think this way, we imprison trees with an urge to travel in the forested areas we have set aside for them. We have completely stifled their innate ability to move north to compensate for rising temperatures.

When forest agencies move trees that occur farther south to more northerly latitudes, they are merely helping them leapfrog over many pieces of property to finally arrive where they were headed anyway. But now it gets tricky. How do foresters know which species would have made it this far north without their help? And even if the species did make it this far, how do foresters know if they would stick around? For some species, this is an easy question to answer. North American Douglas firs certainly don't belong in this category. As they haven't even managed to make it to the east coast of North America, I think we can safely say they would never have crossed the Atlantic to western Europe on their own. And the turbocharged paulownia, native to China, would never have ended up in Germany or in North America, where it is considered invasive in many clear-cuts. In principle, the same can be said for all nonnative species. Even the Turkish hazel, native to the Balkans and across northern Turkey to northern Iran, grows too far from central Europe for us to be sure it would have reached this far north in the next few hundred years.

The second lens through which to view the introduction of exotic species is the lens of pests. The new star trees chosen by foresters have, from a purely economic point of view, a further unbeatable advantage: they are supposedly less susceptible to harmful organisms. Fungi and insects, it seems, are not much interested in them and would far rather snack on beeches, oaks, and spruce. That is indeed true. The little pests have their sights set on native trees. They stick to eating the leaves, bark, and wood of the trees they know. That's pretty much the same relationship we have with food.

Introduced trees are usually imported not as little trees but as seeds. Seeds are free of troublesome parasites, so you could say they are "clean." While spruce and pines are being eaten up by armies of insects, imported Douglas firs, red oaks, and Turkish hazels remain in the best of health. It's easy for foresters to assume all is well. But wait. Gradually, things change. Thanks to international trade, increasing numbers of fungal and insect stowaways arrive, and they are more than happy to find their favorite food laid out in large swathes before them.

One of these stowaways is the Douglas fir gall midge. This midge looks small and harmless. It is so tiny that many larvae fit into a single Douglas fir needle. There, they nibble away protected from birds until they finally emerge, pupate, and start the cycle all over again the following winter. This is catastrophic for the Douglas firs because a midge infestation causes them to lose all their needles—and without their needles, the trees starve. And since 2016, this is exactly what we have been seeing with increasing frequency. Take, for example, the forest around the little town of Rheinbach, not far from where I live. The forester there lamented his plight in a daily newspaper in 2018, saying the trees he was most concerned about were the Douglas firs.[4] Remember, just six months later Julia Klöckner was planting Douglas fir to ready forests for climate change.

What about Turkish hazel? These days, this tree is hard to find even in its original homeland. You come across it occasionally in cities; you hardly ever see it in forests. Turkish hazel is heat- and drought-tolerant and has a lot of lovely features. Its wood is hard and rot-resistant, and just like the shrub variety, its nuts are edible, so there's a lot about it to like. However, an interloper has recently begun popping up in Turkish hazel. The birch sawfly seems to have developed a taste for the tree. Their caterpillars munch the leaves of Turkish hazel down to the veins, until it is impossible for the trees to photosynthesize. That isn't a big problem yet, as there are hardly any stands of Turkish hazel in Germany; however, nature is raising a warning finger and pointing where this journey might lead if foresters were to plant more of these attractive trees.[5]

Planting alien species is a bit like spinning a roulette wheel when you've placed all your chips on a single number. Despite this, foresters continue to pursue the idea of moving trees north—and they've come up with another idea.

FORESTERS ARGUE THAT they could take native species such as beeches and oaks and look for individual trees at the southernmost edge of their range that can deal with a hot climate. Beeches can be found as far south as Sicily and as far southeast as the Black Sea. Wouldn't it work to use seeds from these southern populations to grow heat-tolerant seedlings for planting new forests farther north? The trees' offspring should have enough experience with periods of drought, and you don't have to worry about disadvantages for the native ecosystem. They are, after all, the same species. Species of animals and fungi adapted to beeches won't have any problems with them. Quite the opposite. Their ecosystem and main food source would remain the same despite rising temperatures.

There could well be something to the foresters' arguments, but we should be wary of moving trees around, even trees of the same species. The climate is certainly changing, but no one can predict how fast this will happen and what the effects on local ecosystems will be. Think of the past few years, when it's become drier and hotter surprisingly fast. If you set store by southern populations today, you're betting you can predict the weather for the next one hundred to two hundred years.

May 2020 delivered the best argument yet against putting our trust in foresters who think they can predict the future. In the middle of the merry month of May, the mercury dipped to 14 degrees Fahrenheit (minus 10 degrees Celsius) and fresh new growth and leaves froze even on robust oaks. The choice of trees should not, therefore, be made based on average temperatures, but on the extremes that can occur regionally. Trees live for such a long time that even if cold snaps are rare, you still need to take them into account. What use is a heat-loving tree that will die after experiencing a late frost after ten years rather than two? It still dies.

Beeches and oaks from more southerly regions are at a disadvantage in other ways as well: they have no experience with our local conditions. In addition to late frosts, the amount and distribution of precipitation is different from what they are used to. Even the soil and the diversity of microorganisms that live here present newcomers with enormous, yet-to-be-researched challenges. And if you introduce seeds from other populations to Germany, diseases might slip in that are not even on our radar yet.

Are you familiar with the term *dendrovirology*? This is a new branch of science established by Humboldt University of Berlin. Researchers there are investigating whether trees can get something like the flu. Does that sound like an off-the-wall idea to you? Not at all. It makes sense that plants, like us, might get sick after being attacked by viruses. In the case of trees, one of the culprits is not SARS-COV-2 (coronavirus) but EMARaV (European mountain ash ringspot-associated virus). In addition to targeting mountain ash, this newly emerging virus also infects oaks, ash, poplars, and other species. It damages leaves and thus weakens trees.

You might think trees rarely come into contact with one another. If people stayed in one spot, we never would have had the coronavirus pandemic. Trees do come into contact with one another, however, albeit somewhat differently from us—through insects. Insects in search of tasty plant juices fly and crawl from tree to tree and forest to forest. They carry pathogens with them that get passed along with the next bite into a juicy leaf. The scientists in Berlin are researching other new viruses now widespread in deciduous trees in Europe that, along with harmful fungi and bacteria, are contributing to the trees' decline.[6]

We've known for a long time, of course, that plants can get viral diseases. However, foresters don't look beyond fungi and bacteria. In professional forestry, at least, no one seems to think viruses could play as big a role with trees as they do with us. When you move seeds from southern forests north and sow them there, it's conceivable viruses could be introduced as well. Whether they can do any damage up here and, if they can, exactly what that damage might be is yet to be seen—to know this, we would have to thoroughly research them. As you already know, even bacterial biodiversity is a blank on the map we use to navigate the living world. It's easy to imagine how much less knowledgeable we are about viruses, which are so much smaller.

Tree health could also be influenced by many other factors that we just don't know about yet. Take day length, for instance. June days in Sicily, the southernmost point of the beeches' range, are over two hours shorter than in Hamburg. That sounds irrelevant; however, sunlight means sugar and therefore food for the trees. We don't know what the effect would be of releasing seedlings into the wild that are unfamiliar with conditions in the north. It's true that beeches have been migrating from the south since the last ice age. However, this has been a journey of tiny steps over many thousands of years—time enough, therefore, for the beeches to adapt to new conditions. To put it another way, they have had plenty of opportunity to learn along the way. The southern beeches being moved today, however, have learned nothing about these environmental conditions. Who would their teachers have been? And so the new arrivals here must learn through the hard school of life—and passing grades are not guaranteed.

AND SO WHAT'S THE POINT of helping trees migrate north? Right now, we're seeing that the native beeches in Germany are learning, that they are adapting, and that they are passing their experiences on to their offspring. Once again, we see the tension between impatient people with an eye on economic outcomes and the slow reactions of trees, reactions we cannot help but view with empathy. It seems far from clear that trees are willing participants in this process of assisted migration. It's highly doubtful this is something forests want. So far, our local ecosystems seem to be doing a good job of confronting change—as long as they are not too heavily disturbed, or worse, completely wrecked, by foresters.

As nature strives to regulate itself, it doesn't get a single moment of respite. A group with deep pockets has recently begun to support forest restructuring and the costly establishment of new plantations: companies that want to "go green."

14

Good Intentions, Poor Outcomes

PLANTING TREES IS "IN." Happy people smile at us from television advertisements and promotional brochures: they are out in the forest planting trees to help mitigate climate change. Planting means taking action. The act of planting projects hope and sends a message to the next generation. After all, a tree can live for five hundred years, not only storing a large amount of carbon but also enriching the air with oxygen. On top of all that, it provides a home for countless species.

This all sounds good, but in many cases, the idea is not well executed. Here's a typical example from Germany. At the end of 2020, a large chain of home improvement stores advertised in brochures and on television spots that it was going to plant a million trees.[1] That is basically a lovely idea because we can never have too many trees. A million seedlings, depending on how far apart they are planted, amounts to about 250 to 750 acres (1 to 3 square kilometers) of new forest. But will all that planting really create new forests? On closer inspection, what the chain was actually advertising was that it was going to restructure existing plantations with the goal of making the forest more resilient to climate change. In other words, out with the spruce and in with deciduous trees. This could still be something positive as long as it brings a real added benefit for nature.

To shed some light on these newly planted forests, let's see how the advertisements were translated into action. The chain's partner was the German Forest Protection Association (SDW), a recognized environmental protection organization.[2] Anyone who knows that the German Hunting Association (DJV), for example, also describes itself as an environmental organization understands that this label in and of itself doesn't necessarily mean much.

The German Forest Protection Association has a mission statement. The second item reads: "We place forests with their cultural, economic, and ecological functions at the center of our work."[3] Is it a coincidence that ecological function is listed after economic function? Probably not, as many of the environmental organization's initiatives support the public image of state forestry. For example, they organize youth games together. One of the things schoolchildren learn during these games is how good forestry is for the forest.

The association has assumed responsibility for making the retail chain's promise happen, and it contacts interested forest owners to explain what they need to do. The directives read like the forest restructuring program of state forest agencies. They say they allow only "site appropriate" species to be planted, which is a roundabout way of saying that it's fine to plant exotic species.[4] Were they aware that in forestry-speak, you have to specify "native and site appropriate" if what you want is real nature?

And, truly, if you pay for a tree, shouldn't it be planted in a natural space and allowed to grow old in peace? Shouldn't it live in a forest where the trees support one another, create a cool climate, and increase the amount of rain that falls? In short, shouldn't be a part of and contributing to an undisturbed forest preserve?

Reality is somewhat different—and somewhat more brutal. The seedlings end up once again in forests managed for financial gain and, as this is the way these forests work, the little trees are granted, on average, a few decades of life. In most cases, their wood is destined for economic exploitation. The forest ecosystem is not the only thing that suffers under this arrangement. The air we breathe is also shortchanged as the carbon stored in the trees is returned to it sooner or later. All products made from wood are eventually burned, either as garbage or in power plants that burn waste wood.

New little trees are being planted at a frantic rate up and down Germany even without support from commercial interests. The wildfires around Treuenbrietzen in Brandenburg, not far from Berlin, are a good example of the fate of these energy-intensive and expensive planting operations. In the dry summer of 2018, 1,000 acres (4 square kilometers) of pine plantation went up in flames. I really wanted to see the area for myself. Wildfires are rare in Germany because our original native deciduous forests did not burn. It was only with the introduction of large plantations of conifers, whose resin-soaked twigs and needles are highly flammable, that fires became an increasingly volatile problem.

PIERRE IBISCH IS researching areas burned by forest fires. He has teamed up with biologist Jeanette Blumröder to study how these forests regenerate without human intervention. I—along with Jörg Adolph and Daniel Schönauer (the camera crew for the documentary film *The Hidden Life of Trees*)—had an appointment with Ibisch at one of the burn areas. As we walked among the charred trees, we were astonished to see that some of them, even though they were in rough shape, had survived. I had imagined a wildfire area would be completely barren with just a few stumps rising grimly out of the ashes. Instead, the fire had mostly raged through the understory. It had scorched the pines to a height of 16 to 20 feet (5 to 6 meters) without burning them completely. The forest was still standing, but instead of being brown and green, it now was brown and black.

Our boots stirred up ash, increasing the apocalyptic impression. But wait. Here and there a bit of green was peeping out of the burned ground. We squatted down and carefully touched the little pioneers with blackened fingers. They really were trees. Tiny and barely recognizable, here and there the first maple and pine seedlings were sprouting. You'd have to be quite the optimist to believe that these few beacons of hope among a huge, blackened scene of destruction could be the start of a robust new forest. And yet this is how nature works.

A few hundred yards farther on, I got to see a very different way of dealing with burned areas—and I was amazed. Here the forest owner was cutting down all the trees, whether they were dead or alive. On that May visit, we watched a harvesting machine driving along the edge of the huge clear-cut. It was one of those harvesting machines I described earlier that can fell a tree in seconds, strip it of its branches, and cut its trunk into convenient lengths. The machine on the horizon was stripping the limbs from one downed pine after another, leaving nothing but bleak emptiness behind it. The wide tracks covering the area made the dismal scene look even worse. A forester onsite told us that eventually the ground would be plowed up. He told us that they knew from experience that the conditions in Brandenburg made it difficult for the forest to regenerate on its own.

Some of the ground had indeed already been plowed. In sandy furrows scraped clean of humus, the forest company had planted pines barely 4 inches (10 centimeters) tall. Pines? I asked the forester why the same mistake was being made all over again. Well, that's obvious, came the answer. Everyone knows pines are about the only trees that grow in the sandy soils in Brandenburg. Although I was convinced the opposite was true—after all, centuries earlier this area had been mostly covered with ancient beech forests—the only comment I made was that the whole exercise made no sense from an economic point of view. "But it does," my fellow forester countered loudly and quickly. He proceeded to tell me that the plantation would pay dividends over the decades until it was clear-cut in about a hundred years.

I love my compound interest app because it's really helpful in situations like this. In no time at all, I had calculated whether the investment in the forest really would pay off. "Stop," you might say at this point. "You can't calculate the worth of a forest!" For a true forest, I completely agree. But a true forest is one that regenerates on its own—and forests will do this almost everywhere for free and without any help from us. And as we'll see, the regrowth is almost always more ecologically appropriate, although in some cases, it can make sense to plant rare native trees to increase the ecological value of a young forest.

But if you are reforesting an area with the main goal of one day harvesting timber, then from an economic point of view it should be like an investment in real estate, bonds, or gold, and it should pay interest at a similar rate. Most people thought interest rates were at rock bottom in 2020 and 2021, but this was mostly true for money in checking or savings accounts. All other forms of investment offer good returns regardless of the particular time they are made. Adjusted for inflation, stocks, for example, generate returns of over 6 percent over ten years.[5]

Pine plantations are relatively cheap to start—about 4,000 euros per 2.5 acres (per hectare), including clearing the area and plowing the land. It takes about a hundred years before you can cash in the investment. A century is how long it takes before the trees are large enough to be taken to the sawmill. In the meantime, some timber is harvested as the trees are thinned, but the trunks are skinny and not of good quality. The income they provide usually doesn't cover the cost of management and harvest.

So, we start with 4,000 euros, assume an interest rate of 6 percent, and allow the investment to run for a hundred years. As you can imagine, a huge amount will accrue over this long period of time. It all adds up to over 1.3 million euros. The only way a plantation, an ecological desert, can outperform a naturally regenerating forest is through timber sales. If it doesn't do this, you might as well invest your money in stocks or other investment vehicles. The pine plantation generates a return of only 12,000 euros after one hundred years,[6] making it the worst performer by far when measured against all other forms of investment.

You can make similar calculations for almost all expenses in the business of forestry. The results are clear. Those who work against nature will never achieve a reasonable return on their investment. In short, if you plant, you lose.

WELL-INTENTIONED TREE-PLANTING SCHEMES carried out by companies and private individuals in public forests highlight another unfortunate reality—one that can be laid at the feet of the government agencies responsible for forest management. By planting huge areas with spruce and pines, these agencies have been paving the way for an ecological disaster for decades. Their efforts have been so successful that today more than half the forests in Germany consist of nothing but nonnative conifers.

This has never made sense. Even before the dry summers of 2018 to 2020, more than half the spruce—the most important species for forestry—were falling victim to bark beetles or storms. It was almost as though the industry was designed to fail in spectacular fashion, a predictable disaster in an industry bailed out time and again with tax dollars. In Germany, when disaster strikes, both private forest owners and, of course, public forest agencies have a legal obligation to replant trees within a few years. This means that areas planted by well-meaning individuals in their free time or by generous home improvement chains thanks to receipts at the cash register would have been replanted with new little trees anyway. The authorities in charge, the federal forest agencies, would have made sure of that.

And volunteer initiatives often don't even save forest owners money, because both private and community reforestation initiatives after catastrophes and the replanting of plantations are heavily subsidized.

All that remains is a gigantic public-relations operation. Volunteers are under the illusion they have done something good—and nature bears the costs as it steps back to make way for yet another plantation.

THERE ARE A FEW SITUATIONS where planting a forest from scratch seems like a good idea. It could work, for instance, in places where there are no old trees nearby that could seed themselves—in wide-open expanses of agricultural land, for example. In places like this, natural reforestation would take longer, which isn't a problem in and of itself. After all, there are natural processes at work here and it would be fine to allow them to proceed somewhat more slowly. Nature has plenty of time; people usually have less time. If we view reforestation not only as a return to nature but also, and most importantly, as an initiative to combat climate change, then it's okay to give nature a hand. And in many cases, planting works, especially if you plant native species, which in Germany are deciduous trees such as beeches, oaks, and birches.

Even then, the little deciduous trees face severe challenges from the beginning. The first issue concerns their roots. In nature, a 16-inch (40-centimeter) -tall beech has a root system that stretches out over more than 10 square feet (1 square meter). You can't dig the little tree up and replant it without damaging its roots. And even if you could, the weight of the soil clinging to the roots would require an excavator to move the sapling and put it in the ground. No one wants to go to such an expense to move a small tree—the process needs to be cheap and quick. Just as in agriculture, there is a race to the bottom with prices, and the result is that a small beech or oak must cost no more than 2.50 euros—including the costs of planting. And there you have it, a little tree as large as possible for as little money as possible that can be planted cheaply and quickly.

Let's see where this gets us. As the root system must be as small as possible so it fits into a hastily dug planting hole, the roots are trimmed back at the tree nursery and often once again in the forest. Ouch! Root tips are among the most sensitive organs in a tree. This is where scientists have discovered structures that act a bit like brains. This is where a tree decides how much to drink, which neighbor it will supply with sugar solution via the trees' underground network, and which fungi to pair up with.

Once the tips are trimmed, these sensitive organs never regenerate quite as they were. The little trees no longer sink their roots deep into the ground and barely network with each other. Communication—at least via their root systems—mostly stops, which leaves the infant trees susceptible to dangerous attacks by insects or herbivores. Normally, when a tree is attacked, it uses a chemical alarm call to warn others of its species so they can prepare to defend themselves by building up their stores of toxins. It's as though the young forest falls silent and loses its sense of direction.

Root amputation also leads to flat root systems that no longer adequately anchor trees. When the roots of native deciduous trees grow out rather than down, they no longer have access to the deeper layers of soil where the live-giving rains of winter are stored. These defects are clearly visible in areas of Germany that were newly planted in 2019 and 2020. The trees here often dried up in their first year. In contrast, wild trees of the same age growing naturally in the immediate vicinity were resplendent in their fresh green growth despite the intense summer heat. These wild seedlings have another advantage: they are locals. They know the climate and are adapted to the harsh realities of life in this area. The cohorts from tree nurseries, I am sorry to say, are little softies. They did not need to toughen up during the dry summer because the beds in nurseries are, of course, irrigated. How could they learn to manage their water intake while they were living there?

And then there are the liberal applications of fertilizer in the tree nurseries, which is basically a form of plant doping. The little seedlings in the nurseries get as much food as they want and soil moisture levels that even the best forests cannot always offer. In their first two or three years of life, they are living the dream. The dream ends when the small oaks and beeches suddenly find themselves out in a clear-cut where spruce once grew. They get a rude awakening when their roots are quickly stuffed into a planting hole and tamped down. As planting has left their roots bent and squashed, they find it much more difficult to suck water from the soil—that is, if there's any water available after the ground has been driven over by heavy machinery.

An additional and much more serious disadvantage is that the little trees from the tree nursery don't know enough. They are not only missing what they could have learned for themselves but also what their parents would have passed down to them under normal circumstances. Tree parents pass the distillation of their life experience to their offspring through epigenetic markers. These are methyl molecules that attach to genes like bookmarks and are passed down to the next generation when seeds are formed. The seedlings then "know" right away how to deal with the soil where they are growing, the amount of rain that falls, and the current summer temperatures.

This knowledge, however, really holds only for the few square yards where the tree parents are growing. Think back to those beeches near my home in Wershofen, the ones that react very differently on south-facing and on north-facing slopes. A different set of behavioral guidelines is handed down to—or, it might be better to say, rooted in—each group of trees according to their circumstances. The knowledge stored within the planted-out trees from the tree nursery is of little use to them in their new location. Their mother trees, flawless individuals raised in a few officially recognized stands, have been selected mostly for how suitable the wood they produce is for sawmills. These stands might grow in the Black Forest in southern Germany, while their offspring could be planted in the Eifel in the central part of the country.

Forests that regenerate naturally have one more thing going for them: they are better placed to face new challenges. A beech produces an average of almost two million seeds over its lifetime, all of which have different characteristics. Statistically speaking, only one of the seedlings will grow to maturity to replace its mother. And that one is, logically enough, the one most suited to the exact conditions it encounters where it's growing.

AS MENTIONED ABOVE, the situation is different when planting new forests on former agricultural lands. What is the most effective way to restore these ecological deserts when you want a forest to grow as quickly as possible? It's very simple. All you need to do is simulate natural reforestation but speed the process up. You start by planting the fields with birches and aspen. These pioneer trees are among the first to settle new areas, but if there are no mother trees close by to provide seed naturally, you can help by planting these trees at a density of five hundred per 2.5 acres (five hundred trees per hectare). Growing as fast as 40 inches (1 meter) a year, they will quickly form a small woodland to shade the soil and keep it moist.

Beeches will feel at home in the microclimate under these foster parents, and you can plant them a few years later. Even better, fasten seed trays to posts and fill them with beechnuts or acorns. Jays and crows will pick them up and hide them round and about as winter food caches. Because the birds like to play it safe, they hide up to ten thousand seeds a year, eating only about two thousand of them. Most of the rest germinate in spring—what an economical way of planting trees! All it costs is a few euros, and the seedlings grow with intact roots almost as though they were in a natural forest. "Almost" because they have no mother trees. The alien birches provide shade and moisture in the mother trees' place, but they do not transfer food or information.

THE MORE YOU work with nature, the less spectacular the results. Who can pat themselves on the back when the only part they play is to wait and leave things alone? Media photo ops of people standing around with their hands in their pockets watching nature because there's nothing for them to do aren't particularly impressive. You can only demonstrate that you are ready to jump into action by jumping into action, and you can only show that you are ready to take political action by offering funding.

Our expensive attachment to plantations has consequences not only for trees, but also for animals. And these animals are increasingly coming into the crosshairs of economic interests—and I mean that literally.

15

The New Bark Beetle?

PLANTATION RENOVATION IN GERMANY is being accompanied by a thunderous racket. More and more shots are being fired over the freshly planted little trees. The targets of all this shooting are large mammals, mainly red deer and roe deer. Deer have been labeled as enemies of the fledgling forests and have taken over from bark beetles as the embodiment of evil. The beetles left dying spruce and the hungry maws of wild game now threaten to destroy the young nursery-grown trees planted to replace them. But even here—as you've probably already guessed—a game of smoke and mirrors is being played.

People are looking for a scapegoat. Politicians stand shoulder to shoulder with some conservation organizations in efforts to ease an array of hunting laws and significantly increase the shooting. And they have a point. In many places, deer browsing on the new growth of freshly planted deciduous trees could lead to the widespread failure of this attempt at ecological reforestation. But are deer really the problem?

A natural deciduous forest doesn't support many herbivores. Barely anything grows under the shady crowns of beeches and oaks, guaranteeing hunger. In earlier times, red deer preferred riparian forests where ice carried downriver during snowmelt kept large areas free of trees. The thick frozen chunks that came shooting down on the raging waters of the first floods of the season mercilessly mowed down smaller trees and heavily damaged larger ones. Even today along the Elbe, you can see sturdy oaks that sport scars dating back to the middle of the last century. Once the floodwaters receded, grasses and herbs spread over the riverbanks and riverine meadows. Here, wild cattle, red deer, and wild horses found the conditions to their liking. In summer, when temperatures got too warm and mosquitoes too bothersome, the animals migrated up to the tree line in the mountains. It was lovely and cool up there, and juicy grasses and herbs grew in the open stands of trees.

Roe deer, in contrast, are territorial and stay in one spot. They sought out small, disturbed areas in the forest. When a summer tornado uprooted dozens of old trees or a mighty beech weakened by age toppled over, a small island of light would appear on the forest floor. The sun heated up the humus layer and grasses and herbs found a rare opportunity to take root, at least for a while. Roe deer lived off these scattered patches of green—until humans entered the picture.

Humans changed and continue to change the landscape so much that today forests are nothing but islands of light. Every time a forest is thinned, it's as though a summer tornado has come through and uprooted many trees. Warmth and sunlight now reach the ground. You can see the effect up and down Germany. In the eternal shade of reasonably intact old deciduous forests, even in summer, the understory is mostly the brown color of fallen leaves, whereas the ground under trees in thinned commercial forests is exuberantly green. Blackberries and raspberries, grasses, and shrubby hazels, along with all kinds of other plants that would have had hardly any chance of growing in an untouched deciduous forest, are now running riot.

Roe deer and red deer literally have a field day—or, more accurately, field days—in the new growth covering the ground. They stuff their stomachs with these once-rare delicacies, which, thanks to modern forestry practices, have become boring everyday fare. Evergreen plants such as blackberries play a particularly important role for the herbivore population. Forests usually offer so little food in February and March that many animals starve to death. This sounds brutal but it's nature's way of regulating population density by bringing numbers back down to the level that matches food availability.

This regulating mechanism varies. For example, when beeches and oaks are flush with fruit in fall, a good number of animals can be certain to survive the winter, as the oil- and carbohydrate-rich seeds offer the animals sufficient calories to tide them over until the following spring. In years without beechnuts and acorns, the cupboard is bare. Thanks to modern forestry, in combination with winter feeding by hunters, these times of slim pickings have become rare. Many forests are now also overrun with blackberry bushes, which offer a convenient snack of green leaves year-round, even when snow lies thick on the ground. And so a high population density of wild game exacerbates the problem of failing conifer plantations.

MOST FORESTERS CLEAR-CUT the dead forests, and the ground rapidly heats up in the sunlight. Fungi and bacteria kick into high gear and break down woody debris, needles, and humus in no time at all. Within a few years, such enormous quantities of nitrogen and other nutrients are set free that herbs and bushes positively explode with growth. Plants fertilized to this extent are now particularly nutritious, which immediately attracts roe deer and red deer. The animals process the feast into offspring, and their populations sharply increase.

The deciduous trees from tree nurseries exacerbate the situation. The small beeches and oaks were spoiled in the nursery beds, where they were artificially fertilized and watered. And now they arrive full of juicy, fat buds in clearcuts that have often been completed cleared of any plant life to make it easier for workers to plant the new trees. Perfect. Now it's also easier for the roe deer and red deer to walk around, making their way down the straight rows of delicious deciduous snacks at a leisurely pace, nibbling on each little tree in turn.

I witnessed this dance between wild game populations and forest clearings many times over the course of my career as a forester—although in those days, most of the clearings were created by severe storms. Whether it was the winter storms Vivian and Wiebke blowing through in 1990 or Lothar in 1999 or Kyrill in 2007, each time huge clearings were created that filled with lush green in the following years.

What is surprising is that in the first few years, the browsing by roe deer and red deer on the newly planted young trees is not that noticeable. That's understandable when you consider that the amount of food on offer has massively increased. From the point of view of wild game, when trees fall or are felled, the area available for browsing and grazing increases enormously. It takes time before the animals react to this situation by producing lots of offspring, and so, for a while, they live in a land of plenty. When there's so much food around, young deciduous trees are less likely to be found and eaten.

After a few years, however, the situation flips. In many areas, as the young deciduous trees grow taller, they suppress the green growth beneath them. Alongside the deciduous trees, most of this new growth comes from conifers, which is hardly surprising as the earlier forest consisted almost exclusively of spruce and pines. Enormous quantities of seed lie in the ground, which can now germinate in the aftermath of bark beetle depredations and the death of the older generation of trees, ready to take over most of the future forest. Neither roe deer nor red deer are interested in the conifer seedlings with their resin and essential oils. Open spaces and the food they offer continue to shrink as the young forest grows, and the greatly increased deer population is starting to go hungry. Now every available plant is given an experimental nibble—and every deciduous tree that has been planted is found and bitten off. And so it makes sense to shoot large numbers of roe deer and red deer to protect the newly planted forest, right?

I believe that the argument presented to the public, which follows the steps I have just outlined, ignores a significant contributing factor: the aforementioned clearings made by foresters when they fell trees in reasonably intact forests. Ten years ago, students investigated the results of this practice in the forest I manage. They established that a beech preserve where a particularly large number of old trees were growing was barely damaged by deer. In the shade of old beeches and oaks, the trees' offspring grow extremely slowly and are therefore especially vulnerable. It can take a hundred years for their growing tips to extend beyond the reach of a roe deer's mouth. But, deprived of sunlight, the leaves of the beech children are tough and bitter and therefore unappetizing. Roe deer and red deer avoid areas that do not interest them and most of the young beeches survive.

Just a stone's throw away, in a clearing created by a storm, deer browse on young deciduous trees was considerably more noticeable. The clearing had become something like a twenty-four-hour wild-game restaurant, where many deer dropped in at all times of the day and night to eat their fill. The reason for so-called deer browse—the deer damage on small trees that causes such problems for forestry—is grounded in forestry itself. But foresters lay the blame on the deer—adding to climate change a second cause for forest decline that has nothing to do with the way forests are managed.

I MYSELF ONCE PROMOTED sharply increasing the numbers of roe deer, red deer, and wild boar that could be shot—and made sure this happened. Does that shock you? I stopped hunting years ago, however, based on new research and my own observations.

Back then, I was worried about forests. I wanted to get rid of spruce plantations in the forest I managed and return the space to reasonably natural deciduous woodlands. For that to happen, young deciduous trees had to grow back and mature. My efforts, however, always ended up in the stomachs of herbivores. After tough negotiations, the hunts I supervised raised their kill rates to more than twenty deer per 250 acres (twenty deer per square kilometer) of forest, and the hunters maintained this number for many years. That is more than double the average in Germany, and the small beeches could now grow up with minimal damage from deer. But then along came the students' investigations and with them my part in the disastrous browsing. Hadn't I, a forester who thinned the forest, contributed to the increased number of deer? For months, I pondered ways to bring the forest and the animals back into balance.

I made the following calculation. If hunters were shooting more than twenty deer a year for every 250 acres (per square kilometer) in the part of my forest where hunting was allowed, there had to be at least forty deer there all the time, at least half of them female. That was the only way at least twenty fawns could be born each spring and twenty animals shot every year for years. If the core population were any smaller, it would soon fall apart at this kill rate. And that was clearly not happening.

It's safe to say that the habitat my home in the Eifel offers roe deer and red deer is of average quality, so the situation in my forest can be considered typical. If there are about forty deer for every 250 acres (per square kilometer) of forest across Germany and only about ten fawns are shot, where are the other ten? If hunters really are regulating numbers, there should be more and more deer running around because so few of them are being killed by hunters. And yet, as you can see every time you take a walk in the woods, this is not the case.

During the coronavirus pandemic, we learned about exponential growth the hard way. Sooner or later, this kind of unchecked growth should lead to deer stepping on one another's toes. And yet if you are out walking in a German forest, you're unlikely to catch sight of any large wild animal. So back to the question. Where are the ten fawns for every 250 acres (per square kilometer) that are not shot? There's an easy answer. They die of natural causes just as they have done for millions of years. Hunger, disease, and predators finish them off. The predators that decimate fawns include wild boar and, less often, wolves. In spring, the bristly pigs systematically search meadows with their sensitive noses to track down camouflaged fawns cowering in the grass and eat them up.

INTERESTINGLY ENOUGH, the German public agrees that it's necessary to regulate roe deer, red deer, and wild boar populations by shooting the animals. Why is it precisely these three species where natural population regulation supposedly no longer works? No one is calling for intervention in the case of blackbirds, earthworms, or squirrels. No, these three just happen to be species that have been shot for centuries purely for the pleasure of the hunt.

People have been trying for many decades to control the problem of deer browse on young trees by constantly increasing the number of animals shot. For example, about sixty thousand deer were killed annually in the 1970s. Today, that number is over a million. In the case of wild boar, which damage forests when they consume beechnuts and agricultural fields when they eat crops, the rates have increased tenfold in the same period.[1] Despite the increasing kill rates, the problem of wild game damaging planted deciduous trees has yet to be solved.

There's another reason damage has not diminished despite more animals being shot: the animals are being driven into exactly those areas where they're not wanted. Here's a typical example. An elevated hunting blind sits in a clearing in the forest. The blind has been placed there so hunters have a clear shot. The lack of obstruction is important for two reasons: first, so they can clearly see which animal steps out of the trees, and second, so there aren't any branches that might deflect the shot from its target. No trees grow in this clearing, of course, only grasses and herbs—the ideal food for wild game.

Roe deer and red deer want to get this food, but they know a hunter could be lying in wait. Some savvy deer even take a look to see if there's a hunter in the blind armed and ready to shoot before they step into the clearing. If they're not sure, they wait in the dense trees until dusk, that is to say, until the predatory human can no longer see anything.

The clearing holds both death and promise. And because herbivores must eat almost constantly, during the day they try—and fail—to find something to tide them over in the forest. Instead of grazing on grasses and herbs, they end up browsing on the new shoots and twigs on small trees. They even peel the bark from the tiny trunks. The more intensively deer are hunted, the less they dare to step out into forest clearings and meadows, and therefore the greater the damage.

Despite decades of extensive and unsuccessful effort, the official strategy is still that more animals must be shot. I'd like to remind you here of the definition of madness: doing the same thing over and over again and expecting a different outcome. The take-home message is that the number of deer has spiraled upward because of forestry and feeding. It is the availability of food that regulates numbers, not hunting.

THE FACT THAT NATURE still regulates populations raises a big question about hunting: isn't it basically unnecessary? I know this sentence will horrify many ecologically minded foresters because their worldview includes diversified forests with native trees that they believe can become a reality only if large numbers of animals are shot. Even big conservation groups are caught in this situation, and they too are calling for more intense hunting. But as this strategy is clearly not working—despite increasing numbers of animals shot, levels of damage remain high—it's time to try something new.

I don't know if abstaining from hunting would improve the situation, but we should at least give it a try. We could declare one, or even better two, adjoining districts to be no-hunting zones on a trial basis. We need a lot of space, because if the protected areas are too small, we'll effectively end up with hideaways where animals can escape hunters and then cause even more damage because they're concentrated in one spot. If the no-hunting zone is large enough, a natural equilibrium should be reached—if the idea works.

At the same time, the second significant contributing factor to ballooning populations should be removed: feeding. People are still carting tons of feed into the forest to keep roe deer, red deer, and wild boar in their hunting territories. Many hunters will argue that they don't do this, because supplemental feeding has been illegal for a long time. But that is not completely true—as you can see if you glance at the rules that apply when snow covers the ground (feeding is allowed then). Also, supplemental feeding now goes by another name: stocking bait stations. And so, officially, what we are talking about is feed used to attract game so it can be shot.

Our wild animals have become so afraid of people that just about the only way to shoot them is to trick them in this way. But the bait stations are so well provisioned that the numerous offspring that wild boar and other wild game end up producing are too much for the hunters to handle. If hunting were stopped, bait stations would become a thing of the past.

There's no question that forest plantings are being damaged, and deer are eating the bark even from mature trees. The damage leads to reduced growth and most importantly to noticeably lower-quality timber. And this brings us to the heart of the matter. What we are really talking about when we discuss large herbivores is economic damage and not the damage they inflict on nature. The damage is identified as damage to nature to ensure the public is more accepting of widespread hunting. While we are all happy to protest the hunting of large marine mammals, the shooting of millions of large, beautiful land mammals barely registers.

What is needed now is becoming increasingly obvious: commercial forestry needs to step on the brakes in all remaining forests. Beeches, oaks, and other species have learned over millions of years how to keep themselves out of the greedy mouths of herbivores. Their ability to do this is being severely restricted by our intensive use of wood and by the switch to commercial plantations, which are more susceptible to damage by insects, storms, and drought. Yet this is still something most people do not grasp. And if shooting deer can't fix the forests' problems, then maybe an ancient forest protector can—the wolf.

16

Wolves as Climate Guardians

I'LL ADMIT IT. It seems a little far-fetched to describe the wolf as both a hero in the battle against climate change and an icon of species preservation. I make no secret of the fact that I regard the wolf as an important part of the natural world, and I'm thrilled that it is being so successful at resettling parts of Europe it once called home. These gray-coated hunters are not being reintroduced; they are migrating back of their own accord to places where they once lived. They have been doing this since 1990, when they were declared a protected species. The role of politicians and the public has basically been a passive one. All people have done is to stand back and allow the wolves to return.

The last wolves in the Eifel, where I live, were killed at the end of the nineteenth century, around the time these predators disappeared from Germany. The starting pistol for their return rang out in 2000 when a pair of wolves in Saxony gave birth to the first generation of wolves born in this country in over one hundred years. From there, the species spread ever westward. At the same time, wolves from southern Europe began slowing resettling parts of southern Germany. And even if up until now wolves have rarely been sighted in the sparsely populated Eifel, the numbers across the country show the progress the wolves are making. At the end of 2020, there were 128 packs, 35 pairs, and 10 lone wolves in Germany, meaning there were 173 territories, and in these territories, in the spring of 2020, 431 pups were born.[1]

Wolves eat mainly meat in the form of roe deer, red deer, wild boar, and, occasionally, livestock. Attacks on the latter often make the headlines and bring the wolves into bad repute—unjustly. According to studies undertaken by the Senckenberg Research Institute in Görlitz, less than 1 percent of the wolves' prey consists of livestock.[2] I don't want to get into a debate here about people who raise sheep or other animals—I've already done that in previous books. The much more pertinent question is the data researchers are using to argue that wolves can help us deal with climate change.

The most obvious answer is simple. Wolves eat other animals and mostly large herbivores. Roe deer and red deer, which are over 75 percent of the wolves' prey in Germany, are strict vegetarians. Deer digest the plants they eat. In other words, their bodies convert appreciable amounts of the greens they chew back into carbon dioxide and water. When large herbivores are around, therefore, less carbon accumulates in living or dead vegetation.

It's doubtful wolves have a significant effect on the numbers of roe deer and red deer. To do that—as a quick calculation shows—they would have to eat more than they are physically capable of eating. The average wolf territory in central Europe is between 40 and 135 square miles (100 and 350 square kilometers), depending on the density of the prey population.[3] Let's take the smaller area, that's to say, 40 square miles (100 square kilometers). In territories with a lot of forest, depending on the quality of the habitat, you would expect from twenty to seventy large herbivores (roe deer, red deer, and wild boar) to be bustling about every half a square mile (twenty to seventy large herbivores per square kilometer). That adds up to between two thousand and seven thousand animals in the territory as a whole. By a conservative estimate, these potential prey animals give birth to approximately two thousand to three thousand young every year.

Even using conservative numbers, it quickly becomes clear that a wolf pack would have to bring down quite a few prey animals every day to decimate the population—but wolves don't eat nearly that much. According to research done in the Polish primeval forest Białowieża, the proportion of animals killed by wolves—based on the animals' spring populations—is 12 percent for red deer, 6 percent for wild boar, and only 3 percent for red deer.[4] Compare this with the reproduction rate of roe deer, which stands at about 50 percent. Despite this, wolves clearly do influence population numbers. But how?

Let's approach this question from a completely different direction: by determining the plant biomass of whole continents under a variety of different conditions. This is exactly what a research team led by Selwyn Hoeks from Radboud University in Nijmegen studied. The scientists used computer models to simulate how landscapes change when large predators with a body weight over 46 pounds (21 kilograms) disappear. The result: herbivore populations rise, and plant biomass declines significantly. To sum up the science and relate the results to greenhouse gases, this means that the ability of ecosystems to store greenhouse gases is significantly reduced when there are no large predators around.

THE WOLF IS a prominent example in Germany, but by far not the only member of the club of heavyweight predators. Lynx and brown bears would love to round out this trio if only they could. There are at least a few areas, such as the Bavarian Forest and the Harz Mountains, where the big cats with tufted ears are on the prowl. In general, though, they are a rarity, and so lynx cannot influence populations of wild game—and brown bears, which are nonexistent in Germany, certainly can't. And even wolves are a long way from settling all the territories available to them. Until the day large predators once again play a significant role over a wide area, computer modeling will have to do. And what the researchers discovered is quite something!

According to the models, removing large predators leads to significant changes in their respective ecosystems. One is the increased incidence of diseases that keep the number of roe deer, red deer, and other large herbivores low. The more the animals come in contact with one another, the faster the pathogens spread, something our own species has unfortunately experienced with the coronavirus. The biomass of plants is also lowered, and that's not all. Ecosystems, which are suffering already in these times of climate change, also become unstable when large predators are absent.

The number of smaller predators such as coyotes and foxes increases, which is hardly surprising. After all, they are usually hunted by wolves and other large predators. Without wolves, large omnivores such as bears have a hard time. Their numbers drop in lockstep with declining wolf numbers. The authors of the study attribute this to the fact that the increasing army of smaller carnivores competes with bears for meat (such as meat left on carcasses). At the same time, the massively increasing population of large herbivores decimates the bears' plant-based staples. The effects are less pronounced in areas with large seasonal fluctuations, such as central Europe. Here, winter is a mitigating factor. At that time of year, there is little plant matter to eat, so the number of herbivores cannot expand beyond a certain threshold.[5]

And now forestry once again enters the picture. Blackberries and other attractive winter food, which are now available owing to the massive thinning of commercial forests, cause the numbers of wild game animals to rise way above their natural levels. In such circumstances, even the return of the wolf can no longer restore the natural balance. But it also becomes clear that reducing logging, expanding the amount of forested land, and discontinuing supplemental feeding by hunters would be even more effective if large predators were allowed to play their part.

If that really does come to pass one day (and nothing suggests that it won't), then hunting will not only be unnecessary but, in practical terms, no longer even possible. Natural densities of wild game are only a tenth of the density we are seeing today in our manipulated landscapes. And if natural forests returned, hunters would no longer have targets to shoot at because the visibility of wild game would sink even faster than the population density.

WE CAN COME at protecting forests from whatever direction we want, but the results are always the same. We must exert less pressure on nature by exploiting it less. We must strengthen forests by allowing them to take care of themselves. And we must reduce both forestry and hunting.

The current political solution for climate protection, however, is to use even more wood.

17

Is Wood as Eco-Friendly as We Think?

FOR A LONG TIME, we've thought of wood as an ecologically friendly raw material. When you cut down a tree and burn it, carbon dioxide is released. However, sustainable forest practices ensure that the old tree is replaced by a new one. The new tree grows larger and larger, reabsorbing the greenhouse gases that were released when its predecessor was burned—a classic cycle. This makes wood an almost climate-neutral fuel, as state forest agencies, along with the forest industry, never tire of reminding us.[1]

The official forest users add a further argument: Don't all trees die sooner or later and end up rotting away on the forest floor, of no use to anyone? Rotting means that microorganisms eat the dead tree, and as they do so, they breathe out all the carbon dioxide the giant captured during its long life. Therefore, it makes no difference to the climate whether you burn the tree as fuel or leave it in the forest for the decomposers. Consequently, you can always cut trees down when they are large enough to be of use and plant appropriate seedlings in their place. The circle of life and death remains intact, and we extract a climate-neutral raw material. In effect, all we are doing is removing material that ends up as waste anyway.

Unfortunately, this calculation is completely incorrect. It makes perfect sense, of course, that a tree cannot release more carbon dioxide when it is burned than it took in while it was growing; however, this carbon dioxide would have remained stored in the tree as carbon had the tree not been cut down. Moreover, the tree would have continued to grow and store more carbon, and as it aged, it would have increased the rate at which it sequestered carbon.

Older trees take up larger amounts of greenhouse gases than younger ones, as you can see for yourself simply by looking at the growth rings after a tree has been cut down. Every year, as the tree increases in girth, it grows a new ring between its bark and its trunk. The width of these new rings doesn't decrease appreciably with age as the trunk steadily increases in diameter. As the trunk expands, the volume of the tree increases exponentially and the amount of carbon it stores increases at the same rate. This cumulative growth pattern holds steady until long after the normal age at which a tree is harvested (which is when it is between 80 and 150 years old). Hans Pretzsch from the Technical University of Munich discovered that beeches and oaks don't slow down until they have surpassed the ripe old age of 450, and even then, the rate of growth slows down only gradually and not by much.[2]

A large tree stores a whole lot more carbon in the form of wood along its 165-foot (50-meter) trunk than could be stored by many thin trees taking up a similar amount of space. But there are hardly any big trees left in the world's forests, not in Europe and not even in Canada. With continuous logging and replanting, the average age of a tree in Germany today is only 77 years.[3] Our native trees can live 500 years or more. This means that before we get to the natural cycle of life and death the forestry sector keeps talking about, we still have about 400 years in which the forest could continue to store greenhouse gases. Every time trees are felled before they reach old age, the process of carbon sequestration is interrupted—and not only that. Forests ravaged by timber harvests can no longer cool themselves as efficiently and create less rain, as we've seen from Pierre Ibisch's research.

Moreover, trees in commercially managed forests will never live that long. They can achieve life spans of 400 to 600 years only if they grow slowly in the shade of mother trees in their youth (and we know their youth can last centuries). This shade is destroyed when mother trees are felled. In full sunlight, the young trees grow rapidly and quickly run out of steam. In commercially managed forests, trees live for only 200 to 250 years even when the chain saws spare them.

SCIENTISTS IN POLLINO NATIONAL PARK in southern Italy (just before the toe of the boot) did studies to find out how long the oldest beeches can live. The park covers an area of almost 770 square miles (2,000 square kilometers), making it one of the largest protected areas in Europe. Among the forests that grow here are primeval beech forests and the oldest specimens of this species. By counting tree rings, the researchers established that the oldest beech, called Michele, is more than 622 years old. As the innermost part of the tree is rotten, the oldest rings are missing. When the researchers included an estimate of these rings, the team concluded Michele could be up to 725 years old.[4] That left even me speechless. The oldest beeches I have ever seen are barely more than 300 years old.

Conditions in Pollino National Park are especially harsh, and in circumstances like these, trees are extremely disciplined and grow slowly, which might explain why they grow to be so old here. And yet we get similar reports about primeval forests from regions where growing conditions are less extreme. I know from conversations with local conservationists in Romania that in a small, inaccessible valley in the Carpathian Mountains you can find beeches that are over 550 years old and still in the prime of life. Beeches should be able to live more than 300 years in central Europe as well, if we leave them in peace and allow them to live life on their own terms. It pains me to think that countries like Germany, which once was the epicenter of the global expansion of primeval beech forests, no longer has a single really old beech tree left in its forests.

But back to carbon sequestration. Once we realize that old trees store carbon for centuries, increasing the rate at which they capture it until they are 450 years old, it's critical that we allow trees to grow old. We cannot allow forestry to weaken trees. In the next chapter, we'll take a look at where we should source the wood we need.

Timber producers also try another tack to explain why using wood can help protect the climate: many wood products last for a long time. When you store carbon in wood houses and furniture, you can simultaneously plant new trees in the forest and these new trees will also store carbon. That way, you can store more greenhouse gases than you could in a forest left to grow naturally. There, dead trees would rot, releasing the carbon they have stored—an argument you are already familiar with. And because every tree will meet this fate one day, a naturally growing forest is caught in a cycle that contributes almost nothing to saving the climate. Based on this argument, we should extract wood from as many forests as we can.

WOOD REALLY IS A WONDERFUL raw material, and I love products made from wood. My desk, for example, is made from an old, dead elm, and you can still see the occasional bore hole left by the beetles that caused its demise. The technique the carpenter used to finish the top left the texture of the tree's growth rings so I can still feel them. I do that every now and then when I'm writing something new and allowing my thoughts to wander. Touching the rings connects me to nature, even though what I'm enjoying are in essence the bones of a dead tree and I'm using them not to help the climate but for my own pleasure.

It is, let's be perfectly clear, impossible to extract raw materials in a way that benefits nature. The only choices we have are between more and less damaging practices. Imagine for a moment that when your local baker sells you bread, she tells you that when you eat this bread, you are helping protect the climate. That sounds rather odd, doesn't it? But this is exactly what forest agencies are telling you when they pitch their wares. Both messages are incorrect—and unnecessary. When we have legitimate needs, it's okay to use wood—if we avoid devastating forest ecosystems. But we overstepped this boundary long ago.

BUT BACK TO THE ARGUMENT that long-lived wood products store carbon better than forests. Even if all wood were processed into durable products, the carbon these products contain would be released back into the atmosphere as carbon dioxide within a couple of decades. Professor Arno Frühwald at the University of Hamburg has pieced together how long wood products actually last. Cheap furniture lasts ten years, books do better at twenty-five years, and wood used in home construction (for example, in roof trusses) last for seventy-five years. The average for all wood products is thirty-three years, which is not particularly long for what is touted as long-term carbon storage.[5] In an untouched forest, carbon dioxide would have been captured in the trees for centuries—and cut and processed wood no longer cools any landscapes, nor does it create more rain.

But it gets better (or, should I say, worse). Most wood isn't destined to be turned into wood products; instead, it is destined to be used as fuel in stoves or biomass power plants. The quantity of wood burned—over 60 million cubic meters—equals the total amount of timber extracted from forests in Germany every year.[6] The same amount again is used for other purposes, such as home construction or making paper. This means that aside from salvaging wood waste and recycling, the only way to make up the missing wood is to import it.

To make matters even worse, Germany is getting ready to follow other European countries and switch large power plants from coal to wood. The operators of the coal-fired power plant in Wilhelmshaven are considering switching to wood pellets. This power plant alone would burn almost three million metric tons[7]—the equivalent of six million trees a year.

As early as 2018, about eight hundred scientists warned the European Parliament against promoting the burning of wood in power plants, arguing that this policy would damage the climate and set a bad example for the rest of the world.[8] Even the federal Thünen Institute, which is part of the traditionally conservative Ministry of Food and Agriculture (remember Julia Klöckner?), has come to similar conclusions: protecting forests and putting a moratorium on extracting wood is best for the climate.[9] No matter, state governments continue to fuel the combustion boom through their forest agencies.

The way the industry extracts wood also contributes indirectly to reduced levels of carbon storage. Now is an especially good time to check this out. When you see a clear-cut, know that up to 50,000 metric tons of carbon dioxide per hectare are escaping into the atmosphere from the ground. The exact amount depends on the species of trees harvested. Clear-cuts as large as the ones you see today are illegal in Germany; however, after beetle infestations or storms leave thousands upon thousands of dying trees, foresters rush to remove the evidence. The reason for the devastation lies in earlier efforts to grow as much wood as quickly as possible for industry by planting fast-growing spruce and pines in easily damaged plantations. It's virtually impossible to store carbon in the long term in forests like this, and it's increasingly likely that in the future, the decision about when to empty these stores will be made by catastrophic events rather than by foresters. Permanent carbon storage in forests and intensive use of wood are mutually exclusive concepts. And that is only half the story.

TO UNDERSTAND THE forest carbon cycle, we need to look to the ground. Processes are playing out here that we do not yet fully understand, and yet they have a unique impact on climate change. Generally speaking, soils are the largest land-based carbon storage systems. They store more than all the plants and the atmosphere combined.[10]

Special conditions prevail on the forest floor: it acts like a huge refrigerator. In the shade of giant trees, it stays relatively cool in summer, which means soil life chugs along at a leisurely pace. The pace is so leisurely that increasing amounts of carbon accumulate in a thick layer of humus. If the trees are felled and removed, the ground loses the protective covering they provide and the soil heats up. Now bacteria and fungi amp up their activity, and together with many other soil-dwelling organisms, they deconstruct the brown gold. Within a few years, most of the precious humus layer disappears and the carbon it stored is released into the atmosphere as carbon dioxide.

This process, which can be traced back to forest management practices, is reflected in the data. On average in Germany, soils in forests that have been thinned contain only 2 to 8 percent humus. Any pasture would be proud to compete with those levels: grasslands, which usually lag far behind forests when it comes to humus, average 4 to 15 percent.[11]

Large trees clearly play a special role when it comes to keeping carbon in the soil, as a team led by Australian researcher Christopher Dean discovered. Old trees literally protect carbon, and they guard it so well that up until now their role has been mostly overlooked. When researchers study carbon in the soil, they usually look between the trees—and that makes sense. It's hard work taking soil samples directly under a tree; it's more convenient to dig in the open ground between trunks. The team was working in a primeval eucalyptus forest. Especially under old trees over

3 feet (1 meter) wide, they discovered about four times as much carbon in the soil as they did in the ground between trees. Their results make it clear that switching from primeval forests to tree plantations with skinnier trees must have led to much higher losses of carbon from the soil than previously thought.[12]

Do these results also apply to other regions, such as the native beech forests in Germany? I think they do. It would come as no surprise to discover that a particularly large amount of carbon is stored in the soil under the old trees. This has been a realm of absolute darkness for centuries, a place where no soil has been eroded and no large animals have churned up the forest floor. Moreover, the mighty trees are rotting from the inside. Fungi and bacteria enter through wounds or dead branches and start their work on the inner parts of the trunk. For the most part, this does not damage the trees, because they no longer have any use for their heartwood. Quite the opposite, in fact. Even though the trees are now as hollow as chimneys, their trunks are still sturdy enough to support their crowns. This process of self-composting releases nutrients stored in the trees' inner wood. The resulting humus contains large quantities of carbon that are now locked away as though in a gigantic vault, protected from heat and erosion. If soils are once again to perform as efficiently as they have in the past and if we are to create a new carbon storage bank to make amends for our sins against the climate, then we need one thing above all others: ancient forests. But you already knew that...

IF WE WANT TO calculate the true costs of using wood, then in addition to carbon storage, we should include the effects on the planet's cooling processes and water cycle. After all, what really interests us about climate change is not so much the total amount of carbon dioxide in the atmosphere, but rather the increases in temperature it causes and changes in rainfall. Forests have a considerable influence on both, and if we allow forests to end up at the sawmill, the effects of tree removal in the surrounding areas will be widespread and immediate. The rise in temperature in clear-cuts on a local level exceeds anything predicted globally for the coming decades even under the worst-case climate models. There is no better place to observe this cause and effect in action than in a clear-cut—and clear-cuts are also where changes in forest management practices can have the most impact.

The loss of the forest's cooling effects will be felt into the future even if the trees grow back. Harvesting equipment weighing a gazillion tons compacts the soil. The tracks of these monsters cut through the forests every 65 feet (20 meters). Their wheels cut tracks 10 to 13 feet (3 to 4 meters) wide, compressing the pores in the soil and squashing most of the creatures living there. In some places, every inch of soil is driven over. The damage in Germany likely covers more than 50 percent of the forested areas—and even after thousands of years, this damage to the soil will still be there. In forests in the Eifel, for example, you can see traces of roads built in Roman times; the soil beneath them is still as hard as concrete. The soil's ability to store water is drastically reduced, and winter rains run off into streams and down the valley, causing flooding instead of seeping into the ground under the trees where it would be available to them in summer. The cooling effect of forests is impaired for the long term because beeches and oaks no longer transpire when water is scarce.

And so increased temperatures over large areas in summer can be traced back, at least in part, to the use of heavy machinery in the woods, which destroys the soil's ability to store water. And this indirect effect on the climate must also be placed at the feet of our use of wood. All this makes wood one of the dirtiest of all raw materials, no matter how beautiful products made from wood are.

THE FOREST INDUSTRY in Germany, naturally, sees things very differently. Forest owners point to the positive environmental impacts forests still provide despite all the destruction. To this end, they are demanding that 5 percent of the income from the 2021 carbon tax introduced as part of the National Emissions Trading System (*Nationales Emissionshandelssystem*, or NEHS) be used to support forest owners. After all, it is their forests that are capturing greenhouse gases and making a significant contribution to climate protection.[13]

Of course, young trees also store carbon and plantations also clean water; however, they do this far less efficiently than the original, naturally growing forests. To sum up: first people damage the ecosystem's ability to store carbon and then they want to get money for doing precisely that? It ought to be forest owners who pay for increasingly robbing forests of their ability to help in the fight against climate change. The most promising instrument to do this is a carbon tax, albeit one used in a very different way from the one forest lobbyists have suggested.

18

It's Time
to Pay Up

A WHIFF OF FLOWER POWER always accompanies methods that tread gently on the land—the suggestions sound attractive but are not taken seriously. I get that all the time when it comes to using horsepower in the forest. In contrast to heavy machines, heavy horses that pull trees out of the forest cause minimal damage to the forest floor. And using horses is not that much more expensive, especially if you factor in the damage the steel monsters inflict on the soil. And yet working with animals is considered a romantic throwback to forests past, whereas harvest machinery steered by joystick and computer is like using a smartphone in the forest—a model of modernity and efficiency.

There's a similar trend when it comes to sequestering carbon: out with nature and in with technology. It's called (an acronym, of course) CCS, which stands for carbon capture and storage. Greenhouse gases are captured and stored at vast effort and expense to keep them out of the atmosphere. According to an article in the *Frankfurter Allgemeine Zeitung* published in January 2021, Elon Musk advertised that he would reward whoever invented the best technological fix with a million-dollar prize.[1] If trees could, they would now shyly raise their hands—or rather, their branches—and say: "We've already invented this, although it was more than 300 million years ago. Does that count?"

TO COMPARE THE SERVICES trees offer with what modern technology can do, let's first take a look at the technology. It hasn't got much further than the experimental stage and it sounds slightly mad. First, carbon dioxide is released to create energy, and then this energy is used to recapture carbon dioxide so it can be disposed of. The process is expensive and, in the final analysis, increases fossil fuel use by up to 40 percent. And then the next problem looms. What to do with the captured carbon?

Most solutions envision underground storage systems, perhaps in deep layers of rock. Scientists estimate, however, that only 65 to 80 percent of the greenhouse gas will stay there; the rest will escape back up to the surface. On its way up, the gas can bring salty groundwater along with it, which damages soil.[2] Apart from that, deep rock layers and groundwater are unique, sensitive ecosystems. If we gas them with carbon dioxide, we have no idea what the consequences will be for the communities that live there. And then there are the enormous costs: projects like Norway's Project Longship, where in a couple of years carbon dioxide will be transported by pipeline to be stored 2.5 miles (4 kilometers) below the ocean floor, are estimated to cost 100 euros a metric ton, the unit universally used to measure carbon storage and the one I will use for carbon calculations.[3]

Trees offer carbon storage with absolutely no risk to the environment, and they throw in all the other services forests provide for free. On average, beeches, oaks, and other species store the equivalent of 10 metric tons of carbon dioxide per hectare per year. Using the costs of the Norwegian project, this means each hectare of forest generates an income of 1,000 euros per year. Compare this with traditional forestry, which is "generating" numbers in the red right now. In better years, earnings barely exceed 50 euros per hectare. Instead of relying on complicated, risky technologies, we could turn to trees, which are standing by as ecologically friendly volunteers—for them, after all, carbon dioxide is basic everyday food.

PERHAPS IT SOUNDS too simple or overly romantic to rely on trees. If we continue to fuel climate change at the same rate as in the past, clearly one day even robust native deciduous forests in Germany and elsewhere will die and release the greenhouses gases they have stored. But if it should come to this, and if we really don't turn things around, this will be only one of many problems—thawing permafrost and melting polar ice caps, anyone?

We really don't want things to go that far, and if we now—finally—choose the right path, throwing our lot in with trees has the added benefit that they can get to work right away, if we let them. In the chapter "What's on Your Plate?" I'll show you how vast areas can be opened up for new forests.

WHAT WOULD IT look like if we were put these ideas into practice? The German 2021 carbon tax on energy generated from fossil fuels is a wonderfully simple, fair tool. Taxes like this can be put into place very quickly. My suggestion is that wood be treated exactly like its dirty relations and taxed based on how much energy it produces. Burning wood is more damaging to the environment than burning coal, and that's before calculating in the cooling effect of natural forests and what they mean for local rainfall. There's no point making a distinction between wood destined for fuel and wood for furniture and home construction. We know that sooner or later wood used for the latter two purposes ends up as waste and will be burned.

That makes the calculation easy: 1 cubic meter of wood represents about 1 metric ton of carbon dioxide and should be taxed in exactly the same way as 1 metric ton of carbon dioxide from coal or oil. That would make wood more expensive and prevent it from being burned in power plants as a cheap, ecologically friendly option to replace to fossil fuels.

Wood has true value for the atmosphere only when it remains in ecosystems in the form of living trees. And here part two of my suggestion comes into play: forest owners who leave their forests undisturbed and forgo harvesting the wood should be compensated and paid the going rate per metric ton of carbon based on the size of their forests. Assuming politicians embrace this model of taxation, what would change for the timber industry and forests?

Wood products would not become much more expensive because most of the associated costs are incurred during processing and not when the wood is still a raw material. Applying the carbon tax to harvested trees would create an incentive to recycle more—waste wood will have already been taxed and therefore would be considerably cheaper. The flip side to this is that burning wood would become considerably more expensive. At a cost of 55 euros per metric ton of carbon dioxide, wood as a fuel, depending on how much it is processed, would cost an average of 50 percent more than it does today and finally lose its advantage over other forms of fuel used for heating. People who occasionally light a fire at home to enjoy a cozy atmosphere while drinking a nice glass of wine can certainly live with directing another euro to a climate levy. Of course, completely converting to wood as one's sole source of heat would no longer be worthwhile.

In forests on the outskirts of town, the tax would immediately make its mark in the form of additional biomass. Would dying conifer plantations still need to be cleared? When the timber market is flooded and the overwhelmed sawmills wave away shipments, forest owners will be able to lean back and relax. They will get 55 euros for every cubic meter of wood that stays in the forest. In the long term, the carbon tax will nudge up to 100 euros a metric ton. It's already reached this level in Sweden, and even in Germany some sectors are calling for the tax to be raised.[4]

Especially for wood, the carbon tax and compensation for forest owners could be even higher because fossil fuels that remain buried underground do not contribute in any way to cooling the planet or to its water cycle. They are simply locked away in rock layers deep in the ground like treasure in a vault: safe and inert. In public discussions about forests' impact on the climate, trees are primarily seen as places to store carbon dioxide. However, increasing numbers of scientists are calling for more focus on how forests contribute to the movement of water around the planet.[5]

In all of this, scant attention is being paid to the countless species that lose their habitat when we exploit forests as a source of raw materials. I find it a real shame that such considerations barely register when political decisions are made.

If we accept that a carbon tax is a tool, we could use to effect change as soon as possible. The question becomes:

How easy is it to implement? Wouldn't a carbon premium require a lot of red tape? Not necessarily, if you don't make the tax any more complicated than it needs to be. What if all forest owners were paid based on the average annual carbon storage per hectare across the country, regardless of whether they owned a deciduous woodland or a spruce plantation? That would, of course, be slightly unfair to those who own a particularly beautiful woodland, but regulations need to be simple and easy to follow or you're guaranteed to end up with one thing above all: too many loopholes. On the flip side, every forest owner would have to pay as soon as they felled trees and emptied their carbon bank. Those who cash in legally and those who do so illegally could easily be monitored by satellite.

Although I'm convinced that a carbon tax would lead to more forests being protected, it cannot come anywhere close to reflecting the true value of forests for humanity. When Boston Consulting Group (one of the largest business consulting groups in the world) calculated just how valuable forests might be, it was not the timber they provide but the services they offer to protect the climate that contributed by far the most to the bottom line. To replace the forests' ecosystems services with technological fixes would cost the global economy $150 trillion. To put that into perspective, all the public companies in the world are worth only $87 trillion.[6]

All this supports a severe reduction in forestry and significant cutbacks in how much wood we consume. But the forest industry is not willing to give up just yet, and right at our moment of greatest need, as the coronavirus pandemic raged, a strange argument was brought to the table: toilet paper.

19

The Toilet Paper Argument

"AND WHERE IS THE WOOD TO COME FROM?" I'm tired of hearing this question, which comes up in every discussion about increased forest protection. This is how the argument goes: if we show more consideration and cut down fewer trees while setting up more protected areas, less wood will be available. This will lead to a rise in timber imports from questionable sources. It would, therefore, be much better to extract as much wood as we can from well-managed German forests and keep the areas devoted to preserves in our country small. And yet, as you've just read, wood is being grown in ecologically questionable conditions even here in Germany.

Economic pressure to increase the scale of logging is growing because hunger for wood is rising unchecked around the world, especially in Germany. Politics is behind this, and for years this hunger has been fueled mostly by forest agencies (which sell wood) and by the federal Ministry of Food and Agriculture. In 2012, the ministry issued a press release that proudly stated the annual per capita use of wood had risen by 20 percent since 1997 to 1.3 cubic meters.[1] That comes to a total of 108 million cubic meters of wood a year for the whole country. Depending on where you get the data from, however, actual consumption far surpasses this estimate and could now be as high as 120 million to 150 million cubic meters. It's impossible to be exact because timber extraction in millions of small, private parcels of woodland is not captured in any systematic way. In addition, it's difficult to trace the flow of raw materials in the economy. Is timber being imported or exported? Is waste wood being burned? Is paper being recycled?

The only thing we know for certain is that prior to the summers of drought, the new wood grown by German forests was meeting about half our demand for wood. We have yet to find out how much new wood grows in our forests today—it will certainly be significantly less than before. At this point the problem becomes obvious: if we continue to fell trees at the rate we have felled them in the past, this will lead, even in the short term, to the complete collapse of many forests.

State forest agencies are legally required to protect forests. Faced with a self-inflicted scarcity of raw materials, they're using questionable arguments to prevent additional forest preserves. Here's a claim I've heard particularly often: if we spare our ancient beech forests, then we will have to import wood, for example, from virgin tropical forests. Does having preserves in Germany really prevent the establishment of preserves in other countries? Not at all. In fact, the opposite is true. Because the much-touted German model of forestry claims using wood is a particularly good way of protecting forests and Germany therefore leaves barely any forested area untouched, other countries follow our example—as we are hearing from countries such as Romania. People wonder why we need forest preserves if the forest outside these refuges is so much healthier. But now that it is becoming increasingly clear even to nonexperts that forests are not, in fact, healthier when trees are cut down, it's time for the industry to play its trump card: toilet paper.

Since the coronavirus pandemic, toilet paper has stepped into the spotlight as the Achilles' heel of modern civilization. It won its right to audition for the role thanks to the panic buying and supply-chain bottlenecks in the spring of 2020. Toilet paper is made from wood fiber, mostly from trees grown in plantations. The main species are spruce, pines, and eucalyptus, but birches and beeches can also be used. The main requirement is that the trees must be cut down and processed. The take-home message is that protecting forests and producing toilet paper are mutually exclusive. When it comes to primal fears, running out of toilet paper appears to be a much scarier prospect than running out of forest.

If you replace wood for toilet paper with wood for—take your pick—construction, furniture, or books (yes, I am also caught up in this), then the situation becomes clear: if we want to do more to spare the forest, then our very lifestyles are threatened. An entire group of professional, state-appointed forest guardians is now appealing to our primal fears because our reason has been telling us for a long time that something's not right. No tactics are too dubious, it seems, when you want to hold on to the old ways of doing things.

However, in its eagerness to promote plantations, the forest industry is completely ignoring the fact that it is the industry itself that is drastically restricting the amount of wood available over the medium term. Once the dying plantations have been harvested and sold at a loss on the timber market, the raucous party will be over. It will be decades or more before the trees can once again be harvested from the now-barren clear-cuts. And since in Germany alone nonnative conifers are cultivated in over 50 percent of the country's forested area, a correspondingly large amount of forest is likely to die off in the next five to ten years. After the overabundance of fallen timber, there will be a wood famine—and with that will come a lot of wailing and gnashing of teeth. If, instead, we allow natural regeneration to happen now, the forests that regrow will be much more robust—and that will benefit the forest industry in the long term.

There's no doubt that we'll need wood in the future. It is our most natural raw material. It's just that, unfortunately, wood is not nearly as ecologically friendly as most people would like to believe. Moreover, today's society is greedy for raw materials, and wood is no longer available in the quantities we demand. We should keep that in mind when we buy furniture, paper, and other products, and we should be more frugal in our use of resources.

A radically different solution to sourcing wood is required. The forest industry wants to make forests match our hunger for raw materials. As this is no longer a viable option, we should turn this around and instead ask how much wood forests can provide, how much we can intervene, and how many trees we can take before we do untold damage to the way this important ecosystem functions.

The answers to these questions are crystal clear: we don't know. All the models assume tree growth is predictable. Foresters have traditionally used what are known as yield tables. Sample plots of different species in different locations are measured for many years. Scientists then compile the results into tables that any forest owner can use to calculate how much wood their own spruce, pines, or deciduous trees will produce annually per hectare.

Once the stands of trees were measured, these yield tables worked for many decades to give a rough idea of how a forest might be expected to grow. But then, at the turn of the last century, it turned out that forests were growing faster than anticipated. The increase was between 10 and 30 percent. The reason for the growth spurt was pollution from traffic and agriculture, which brought nitrogen into forests and fertilized them heavily, damage that continues to this day. Is rapid growth a problem? Yes, it is, because trees by their very nature prefer to grow slowly, carefully portioning out their reserves of energy. They need this energy not only to grow trunks, branches, and leaves, but also to fight off diseases or to pay the fungi in the soil that help them communicate with one another.

Trees used to filter from the air a maximum of 110 pounds of nitrogen compounds a year per 0.4 square miles (50 kilograms of nitrogen compounds per square kilometer)—a tiny amount that had a correspondingly marginal fertilizing effect. Human activities have caused this to climb as high as 11,000 pounds (5,000 kilograms), more than a hundredfold increase.[2]

With all the nitrogen in the air, it's as though the trees are being doped up on performance-enhancing drugs, and they grow beyond sustainable limits. Harvests have increased and the amount of timber dumped on the market has grown steadily over the years. But this dream is now over—the forests are burned out. If nitrogen contamination continues to rise, growth rates will slow down once again as nutrient cycles get out of balance and trees start to apply the brakes.[3]

Whereas the percentage contribution to nitrogen pollution from traffic is lower than it used to be, the portion from agriculture continues to climb—in large part because nitrogen gas released from liquid manure when it is applied to fields then drifts across the landscape. This gas continues to enrich the forest floor with a natural fertilizer that used to be in short supply, affecting the growth not only of trees, but also of plants on the ground. A riot of stinging nettles, black elderberries, and blackberries are proof that these few species are throwing a fertilizer party and living it up at the expense of more frugal species and the trees' offspring.

Add the stress of climate change to overfertilization, and you can basically toss out the old yield tables, because when faced with high temperatures combined with long-lasting drought, trees do more than slow down—they come to a complete standstill for weeks on end. Trees arm themselves against heat and drought by closing the openings on the underside of their leaves or by jettisoning their leaves altogether. In both cases, they lose the ability to photosynthesize, which effectively shuts down wood production. In short, these days, even in intact forests, we can no longer make a reasonable estimate of how much wood will grow there in the future. Anyone who wants to further increase wood consumption is quite simply acting irresponsibly.

AND THAT BRINGS US BACK to toilet paper. Apart from the fact that you really should buy the kind made from recycled paper, this is another area where society has progressed. You can get special toilets (or attachments for your existing toilet) so you can rinse instead of wipe—and some even come with a built-in air dryer. I admit that I haven't tried these bidet systems yet, but when our forests no longer produce enough wood for all our needs, I personally would opt for one and leave paper to be used in books.

But traditional forestry in Germany has not given up yet. When things have gotten sticky in timber markets, it has always been able to count on getting money elsewhere—from the government. The wider the government opens its purse, the better—and right now it's wide open for forests.

20

More Money,
Less Forest

THE FOREST IS DYING a second time. The first time, in the 1980s, acid rain was such a threat to the planet's green lungs that I was fearful for the future. In 1983, I had begun my internship prior to studying forestry, and I wondered if I would ever be able to follow my chosen career. Television documentaries were presenting extremely gloomy pictures of landscapes devoid of trees disappearing into dismal brownish-gray backgrounds. They predicted that by the year 2000 at the latest, large expanses of forest in Europe would be a thing of the past.

The fact that everything turned out differently is not proof that the catastrophic scenario was overblown. Quite the opposite. The frightening reports led to massive political reactions, including, for example, the invention of new technology to remove sulfur from gases released as industrial waste. Catalytic convertors for vehicles also became standard—and forests breathed a sigh of relief. Unfortunately, our collective memory of this unprecedented environmental success is fading. It is imperative that we remember what is possible when forests are threatened—a threat to forests is also a threat to our future.

The second round of forest dying started in 2018. The trees in thousands of square miles of spruce plantations in Germany lost their needles, and in no time at all, people came up with the moniker "Forest Dieback 2.0." Unlike the first dieback, back in the twentieth century, this one is happening much faster and is especially obvious because, despite frantic efforts, it is no longer possible to hide the evidence from the public eye.

But let's start at the beginning. The first forest dieback threatened foresters as well as forests. For despite all the damage being done to trees by growing them in plantations and despite the use of heavy machinery, the immediate cause of the dieback was dirty air from industry and traffic that was burning the trees' leaves and needles. In addition, there was the possibility that acid loads in rainwater were destroying the smallest mineral particles in the soil—clay minerals— which are important for storing nutrients. This damage was something humans would never be able to repair. The image of foresters as protectors of the forest emerged unscathed, and the dramatic situation even solidified it.

Things are different today. The external threat is similar, stemming as it does from widespread changes that are causing forests to die. A striking difference, however, is that the dieback is happening first and most forcefully in plantations, plantations not of native trees but of spruce or pines. Beeches and oaks in forests that have been extensively thinned (or plundered for their wood) are also dying back, whereas deciduous trees growing in large, intact preserves are strikingly resistant.

The comparison between different types of forest highlights that the underlying cause of this weakening of ecosystems is conventional forestry. Climate change is the tipping point for systems that were already teetering on the edge. It doesn't help that some foresters have joined a mini-movement called "Foresters4Future" that is a brazen copy of the youth movement "Fridays for Future." Those who caused the crisis are now advertising it to trigger a sympathetic response from the public.

Strictly speaking, what is dying is not the forest but "only" trees. The ecosystem itself is still functioning, as the burned areas my film crew and I visited in Treuenbrietzen show. In all the places that are now being left alone, the forest is reacting vigorously and new trees are immediately starting to grow. Only where everything has been cleared away, where the soil is warming up under the full force of the summer sun, where the forest floor has been flattened by machinery, and where barely any humus remains is the forest itself actually dying.

And the government handouts now raining down on these clear-cuts will, unfortunately, do nothing to stop those responsible from continuing to do what they are doing. And because in their opinion nature can no longer manage real forests—only forest agencies can do that—it's clear these agencies need financial support, given the enormous areas that had to be clear-cut. And so the question to be put to the public is this: How on earth are we going to tackle the daunting task of once again having healthy and resilient forests?

According to the forest industry, the money from the federal government, which amounted to over 500 million euros in 2020 alone,[1] is only a drop in the bucket and not nearly enough. Is that really the case? Or is it perhaps exactly the opposite, and this money is, in fact, the drop that is causing the bucket to overflow? The large sums now pouring out of government coffers are primarily to promote planting new forests—in other words, to create new plantations. We've already seen from the example of the wildfire areas in Treuen-brietzen that this is not only a losing proposition financially, but the grant money literally goes up in smoke. What all those billions of euros in subsidies are actually doing is prop-ping up a rotten system far removed from nature—a system that would break down without this financial support. And by obstinately holding on to an agricultural model—the idea of trees as crops—this system also ensures that the new forests are increasingly unstable and therefore are not as long-lived.

I think all these new plantings are about something else entirely. Forest agencies are trying, at great expense, to erase all visible evidence of the problems they have caused—an impossible task, as you can imagine, given the size of trees and the vastness of forests. This is less about deceiving the public about how badly a whole sector has failed, and more about dealing with guilt. Who enjoys looking on as their life's work dries up or gets eaten by insects? I have colleagues who gave up and took early retirement in 1990, the year multiple storms uprooted all the trees for thousands of square miles, remaking the contours of the landscape for decades. Back then, the damage also happened overwhelmingly in conifer plantations, which were cleared as quickly as possible and then hastily replanted.

Even if "out of sight, out of mind" doesn't actually work, the idea is comforting. To remove the ugliness you can see—or at least soften its impact—is not so much an official cover-up as a deeply human response. You could interpret government intervention as a gigantic effort to fix things, while knowing full well nature cannot be fixed. The idea behind this intervention is to clear away everything and restart the forest from nothing, to start over with a completely clean slate. To do this, foresters—as we know—are looking for the appropriate supertree so they can go ahead and replant the clear-cuts. Technically, the disaster is then erased and there's forest once again. In no time at all, the issue can be considered resolved.

Here's the problem. Losing—or, more accurately, forcibly removing—all these trees costs the German forest industry a great deal of money. Few sawmills will buy the dead spruce because fungi and insects swiftly spread through the trunks, affecting the color of the wood and leaving ugly holes. Who wants to buy boards, furniture, or even roof trusses made from wood compromised in this way? Little wonder that the price plummets below the cost of felling, processing, and transportation.

Every trunk stripped of its branches and dragged out of the forest is a testament to economic failure. Had the trunk been left in the forest, it would have offered a home to countless tiny creatures; it would have stored water and cooled the air around it. After many decades, it would finally have decomposed into humus and enriched life in the soil for centuries. Looking at the forest from an ecological perspective was and still is completely foreign to the political decision-makers. How else could they fund a huge undertaking to clear "damaged timber," which is how they refer to this valuable biomass?

LET'S LEAVE THE FORESTS and take a look at the inner workings of the forest industry to see what havoc this flood of damaged wood causes. In a normal year, the industry cuts down about 28 million cubic meters of spruce trees in Germany.[2] They are relatively easy to sell, and deducting the costs of harvesting them, the industry clears about 60 euros per cubic meter. Sawmills set great store by wood that is absolutely fresh, because in summer, wood begins to rot after just a few weeks.

According to official estimates, in the three years between 2018 and 2020, 178 million cubic meters of damaged timber piled up, overwhelmingly spruce attacked by fungi or beetles. It's no surprise that prices dropped to rock bottom and have stubbornly stayed there. The cost of harvesting aside, things are getting expensive for some forest owners. Their typical, reflexive response is to scream for financial support, and this flows in a steady stream. Depending on the state and region, the government pays up to 30 euros per cubic meter[3]— which often covers the cost of processing the trunks. And so government provides an incentive to remove biomass from the forest, where it has great value, to the timber market, where it has almost none.

This whole bizarre process does have one positive side effect. Chinese buyers have woken up to the overcapacity in the German timber market. Large trees at bargain-basement prices—it's time to buy! And so thousands of containers leave German ports headed to Asia. During a telephone conversation with Frank Voelker, the administrator for the Kwiakah First Nation in British Columbia, I realized how the processing of damaged timber in Germany is having an impact around the globe. Voelker told me that the chain saws on the First Nation's reserve had been silent for months—no Canadian company could sell timber at such a low price, even from clear-cuts. That meant the coastal forests of the Pacific Northwest could rest easy for a while.

FOREST SUBSIDIES ARE tied to politics, and I offer this example from Germany as a cautionary tale. Most of the federal money does not go to covering the costs of felling beetle-infested trees. The assistance is intended to support reforestation; however, it's paid out in a lump sum without requiring proof of how it's spent. The sums come to 10,000 euros or more for every 250 acres or so (10,000 euros or more per square kilometer)—which means that when it comes to grants and plans gone disastrously wrong, forestry is now on a par with conventional agriculture.[4]

One of the forest industry's most powerful lobby organizations, the Federation of German Forest Owner Associations (AGDW)[5] has been hugely successful in extracting this shower of money from Minister of Food and Agriculture Julia Klöckner. The association's president, Hans-Georg von der Marwitz, is a member of parliament for the Christian Democratic Union (CDU). According to a website that monitors lawmakers (abgeordnentenwatch.de), in 2021 von der Marwitz came second on the hit list of members of parliament with the highest supplemental incomes.[6] The AGDW represents conservative forest practices and in the past has, for example, not shied away from joining farmers' groups to intervene against the ban on insecticides.[7] Regional chapters of the AGDW include not only private individuals but also representatives from community and state forest agencies. In other words, state authorities are exerting a covert influence on the federal government's funding policies through private associations.

The government contracts with another private organization, the Agency for Renewable Resources (FNR), to distribute the funds. The organization was created by the federal government in 1993. One of its tasks is to coordinate and develop grant programs. In addition, the agency is tasked with gathering and making available the latest scientific thinking on renewable resources, including extracting energy from wood.[8] Fascinatingly, for its part, the agency stubbornly ignores that burning wood is extremely harmful to the climate, instead preferring to state that burning wood is carbon neutral[9]—contrary to the opinion of an overwhelming number of scientists. Members of the agency include—this is no surprise—members of the federal Ministry of Food and Agriculture, the timber and forest industries, and other state agencies.[10] In short, I believe what we have here is a kind of cash machine that draws up a list of needs, pushes for a majority vote, and then distributes the money acquired for the benefit of its own members.

None of this helps the forest because the payments come with few strings attached. Take the lax process for the Programme for the Endorsement of Forest Certification (PEFC). This certificate barely covers anything more than what is already required by law; it is cheap, and it imposes hardly any conditions on forest owners.[11] No wonder most forest companies in Germany have wrapped themselves in this mantle of ecological correctness. They are richly rewarded for leaping over a hurdle that is basically lying flat on the ground. Whether the beneficiaries use the cheekily named "sustainability premium" to buy a new car or renovate their living rooms really doesn't matter.[12]

Clearly, it was in someone's interest to avoid a controversial discussion about these funds before parliament voted on them. How else can you explain how this forest premium popped up as small attachment to an "Agriculture Products

School Program" law that was being passed.[13] The idea behind the awkward name was to distribute fruits and vegetables to children in school. Debated in parliament late at night, only two MPs, one from the Greens and the other from the Free Democratic Party (FDP), spoke against it.

Just to be clear, I am absolutely in favor of helping forest owners financially, but we shouldn't repeat the same mistakes we made with agriculture, where substantial farm income comes from grants that are not tied to appropriate environmental standards. Instead, we should be encouraging businesses that are working for the return of robust ecosystems and therefore offer a real return on investment for the global community.

Even without subsidies, money intensifies the forest crisis. The credo of many communities and state forest agencies is that the forest sector should provide substantial income to pay salaries and turn a tidy profit to fill the coffers.

That doesn't work quite as well with wood products as it does with agricultural products because regular catastrophes such as bark beetles and storms upend the timber market. Whenever a lot of damaged timber comes on the market, prices drop, and that causes havoc with the communities' finances. Mostly, it's spruce and pine plantations that are hit by such events, and when trees of this species all fall at once, there are hardly any buyers. This is not problem for creative foresters; they simply move their harvesting equipment over to the old deciduous forests that are often standing strong in the face of climate change. Here, they can find mighty oaks and beeches that still command good prices on the market. It leaves a bitter taste in my mouth to watch the most robust and ecologically valuable forests also being ravaged.

After the substantial clearing that happens when trees are cut down, the old trees left standing suffer in the bright sunlight that now hits their trunks. The delicate beech, with its smooth bark, is well known for its susceptibility to sunburn. The bark splits, exposing the sensitive wood, and fungi and bacteria immediately move in. The giants' fate is sealed and they will die within a few years. Like a smoldering fire, the die-off of deciduous forests creeps along in lockstep with the death of the plantations, with one important difference. Increased summer heat is causing the death of plantations; chain saws are causing the death of ancient deciduous forests. It is, therefore, imperative that logging be banned immediately in all intact deciduous forests.

There are, however, players in the world of science who want to prevent the expansion of protected forest areas, and their methods are deserving of deeper scrutiny.

21

The Ivory Tower
Wobbles

TOBIAS WAS INCENSED. My son was sitting in his office at
the forest academy in front of a huge photograph of beech
forests resplendent in the bright greens of May. The mood
was dampened by figures from a recent scientific paper out
of the Max Planck Institute for Biogeochemistry in Jena that
were flickering across his screen.[1] Up until then, this research
institution had lived up to its august name. The institute has
initiated impressive studies on how plants store carbon that
I often refer to. This time, however, there was clearly something off about the numbers—about the whole paper, in fact.

The paper was written by Professor Emeritus Ernst-
Detlef Schulze, who had taken up his pen to write once again
for his former institute and invited coauthors to join him.
One of the coauthors was Professor Hermann Spellmann,
who at that time (February 2020) was chair of the Scientific
Advisory Board for Forest Policy, part of the federal Ministry
of Food and Agriculture. It's therefore reasonable to assume
that both authors exert considerable influence on the federal
government's forest policies.

The scientists' take-home message was that it's better
for the climate to cut down trees and burn them to produce

energy than it is to protect forests. The federal advisory board supported these findings with an expert opinion of their own.[2] Excuse me? Think of the Amazon rain forest and its importance to the climate not only of South America but also of the whole world. Think of the climate studies carried out at Eberswalde University that describe the enormous cooling effect of ancient deciduous forests. In 2008, Schulze himself had coauthored a much-cited study in the highly regarded scientific publication *Nature* that credited forests with a high potential for storing carbon.[3] And now this!

Schulze stated in the institute's press release: "To achieve the greatest possible protection for the climate, we suggest the planned carbon tax on burning fossil fuels should be used to support the sustainable production of wood." I'll spell this out. It was clearly not enough for him to use his position as a scientist to greenwash the burning of wood as climate friendly; he went even further and asked for money from tax revenues. Can you imagine an oil baron demanding federal funding for burning petroleum? The professor is certainly no oil baron, but the comparison is not that much of a stretch. Schulze is actively involved in forest management in his role as managing director of two forest companies in Germany. I find his work in Romania even more troubling. According to his home page on the Max Planck Institute website, he is the deputy manager of a forest enterprise in that country, as well.[4]

Some of the last primeval beech forests in the world grow in the Carpathian Mountains, and they are suffering the same fate as the Amazon rain forest as they become pawns in the games the forest industry plays. Just like in Germany, all kinds of excuses are made to cut down giant old trees as if there were no tomorrow. Some of the arguments sound similar to those we hear from forest agencies in Germany, Sweden, Poland, and other countries: bark beetle–infested trees must be removed as quickly as possible to prevent the infestation from spreading from hot spots to the whole forest. The salvage operations are called "sanitary felling." These operations become much larger than expected and are often the starting point for further felling in the area.

Roads are bulldozed through the few remaining pristine mountain valleys to give heavy machinery access to the old beech trees. From there, chain saws hack their way left and right up the mountain slopes. The situation is basically no different from what is happening in tropical rain forests, except the clearing is taking place in the middle of the European Union, which supposedly has high standards for environmental protection.

Romania has become one of the leading timber producers in Europe, catering to large companies such as IKEA. Foresters who stand in the way of illegal logging can be brutally murdered, as happened to Răducu Gorcioaia. When he caught tree thieves red-handed, they killed him with an axe.[5]

BACK TO PROFESSOR SCHULZE. According to statements from local environmentalists, he is involved in logging operations in Romania's western Făgăraș Mountains. And here we come full circle with the comparison to an oil baron, as this makes it clear Schulze is a timber producer—especially given his involvement with the two German forest companies as well—and therefore subject to a certain bias. This becomes an issue when you look at the remarkable results of his calculations. Indeed, they are more than remarkable.

Tobias explained to me that Schulze had made a serious mistake. The explanations that follow may not seem particularly interesting and are unfortunately a bit difficult to unravel, but I want to walk you through them because they show the tricks used to create what is allegedly forest science. The controversial aspects of this study, which is critical for how forests are managed in many places and not only in Germany, are a good example of the arrogant attitude of key influencers in forestry.

Among the crucial baseline data were measurements made in Hainich National Park in Thuringia. This small national park in central Germany protects mostly older beech forests. Although these forests were conventionally managed in the past and heavy machinery was driven through them, they are now being left alone to transition to primeval forests. Schulze used the national park as his reference point for forest preserves, which is in and of itself somewhat questionable, as it takes decades or even centuries before forests like this are even halfway restored to a primeval state.

To prove how much (or how little) carbon unmanaged forests store in their biomass, Schulze used measurements taken at 1,200 sample points in the park to determine the amount of wood in the trees in Hainich. In 2000, the average came to 363.5 cubic meters of wood per hectare. When the study was repeated in 2010 using the same sampling points, the amount had risen by an additional 90 cubic meters per hectare—obviously, because the trees had grown over the past ten years. This meant that every hectare that had previously been studied in the national park had added 9 cubic meters of wood per year, which corresponds to about 9 metric tons of carbon dioxide removed from the atmosphere by the forest. This result did not surprise experts because it was roughly in the range recorded for every older beech forest in Germany.

When a second set of measurements was taken in 2010, national park staff sampled additional areas, where either no trees or only very young trees were growing. No problem—to calculate the increase in wood in the old beech forests, all the researchers had to do was to discard these additional measurements. After all, you can't use data that includes newly sampled young forests for purposes of comparison in a scientific study—you can only consider the areas that had been sampled back in 2000. The manager of the national park, Manfred Grossman, made exactly this point in the study. Anyone who then used this data for their calculations could no longer claim to have done so by mistake.[6]

No problem for Schulze, or perhaps even an opportunity. For whatever reason, the professor included values from the young forest—and on paper the average amount of newly acquired wood biomass sank from 9 cubic meters per hectare per year to 0.4 cubic meters. That is less than one-twentieth of the correct amount.[7] Hurray! According to his calculations, old unmanaged forests store barely any carbon, while old commercially managed forests (according to demonstrably correct numbers from the national forest inventory) store about twenty times as much.

Based on this data, Schulze, Spellmann, and their colleagues calculated that intensive forest use leads to higher rates of carbon storage in the forest, which greatly benefits the climate. Hmm. Removing storage units leads to higher storage levels? The forest industry was elated, while environmentalists were appalled. Somebody would pick up on this, right? Tobias met with other scientists, and under the leadership of Torsten Welle from the Natural Forest Academy in Lübeck and Pierre Ibisch from the Eberswalde University for Sustainable Development, a critique was published internationally and a press release was posted on the university's home page to make the world aware of the mistake.[8] Two other teams of international scientists also criticized the study.

The reaction was swift. The Thünen Institute of Forest Ecosystems, which reported to the federal Ministry of Food and Agriculture and then minister of agriculture Julia Klöckner, intervened. The task of this federal research institute is to keep politicians abreast of current scientific knowledge.[9] However, instead of criticizing Schulze's scientifically questionable study, the director, Andreas Bolte, unexpectedly took to Twitter to criticize the scientists who had highlighted the mistake.[10]

The new chair of the German government's Scientific Advisory Board for Forest Policy, Jürgen Bauhus, also weighed in. This board is tasked not only with developing recommendations for politicians, but also, expressly, with promoting scientific discourse.[11] Bauhus, who teaches forest science at the University of Freiburg, has an unusual understanding of the word *discourse*. He issued a written ultimatum demanding that the press release be corrected, as it also criticized the scientific board, which, following Schulze and Spellmann, had spread the word that using forests was better for the climate than protecting them. To top it all off, to apply pressure to the university and, above all, to the participating researchers from Eberswalde, Bauhus complained to the German Research Foundation (DFG). Needless to say, the research foundation could not establish any breach of scientific protocol, and it set aside its investigation.[12]

THE HUBBUB KICKED UP by the study was a seminal moment in my thoughts about forest policy. When you not only criticize people in public institutions who think differently but also try to silence them and damage their careers even though they have done nothing wrong, that worries me. But things got worse.

Schulze and Spellmann's study is more than just a painful misstep for the Max Planck Institute and German forestry. Far more problematic is how this played out in Romania. Schulze enjoys a certain renown in that country, and when a German scientist along with forest experts make a recommendation to no longer protect old forests but to make use of their wood instead—and in so doing destroy them—that's a slap in the face for all the local people working there to protect the forests. Many of them are risking their lives to protect the old beeches for the good of humanity.

Christoph Promberger, executive director of Foundation Conservation Carpathia (FCC),[13] told me the Romanian state forest authority was thrilled to read Schulze's paper, as it legitimizes the country's heavy-handed approach to logging. Promberger tried to purchase the section of forest being logged in the western Făgăraş Mountains and include it as part of the foundation's project to create the largest national wilderness park in Europe. Unfortunately, the environmentalists failed in their efforts to save and protect the old trees.

You might almost think Schulze wrote his paper mostly for himself to justify his participation in one of the largest acts of environmental destruction in Europe. The collateral damage stretches much farther than the few square miles of German and Romanian forests under Schulze's wing. If you're reading these lines in another country, they are just as relevant for you as they are for German readers. It's not only because we all have the global climate in common and therefore depend to some extent on every forest in the world. It's more than that. It's also because German forestry has influenced forest practices worldwide since the nineteenth century and continues to do so because, much to my chagrin, it is still considered to be an exemplary model to follow.

That said, there are definitely experts from other countries who have recognized the harmful impacts of the German model—for instance in the forests of India. Pradip Krishen, one of the most respected environmentalists and nature experts on the subcontinent, wrote in the introduction to the Indian edition of *The Hidden Life of Trees* that it was German foresters who introduced the ideal of same-age plantations to people in India, who then clear-cut forests, planting only desired species and removing all others. This style of forest management has, according to Krishen, caused enormous damage in India, damage from which the system has not yet healed.[14]

NOW COMES THE QUESTION of why people the world over listened so intently to what German foresters had to say. In the nineteenth century, what was then understood as modern industrial forestry was essentially carried out only in France and Germany. Large parts of the world were ruled by Britain, and the British had well-publicized problems with the French. And so it seemed like a good idea to invite German foresters out to the colonies so nature there could be thoroughly tamed. The empire is history but plantations, unfortunately, are not.

The insistence on repeating the mantra that burning wood is climate-friendly reminds me of the oil industry. The Anglo-Dutch oil company Shell knew how harmful their fuels are for the climate from its own research as far back as thirty years ago. And yet it joined other industry giants to deny the climate change they had all caused.[15] The forest industry, too, is denying the scientific consensus that burning wood damages the climate. The scientists suggest that in some cases burning wood is even more damaging than burning black coal. As noted earlier in this book, about eight hundred scientists warned the European Parliament about this as long ago as 2018.[16]

A study from 2017 indicated with concern that with the European Union's goals of switching to renewable energy (which includes wood), the use of wood in the EU would more than double, from 346 million cubic meters in 2009 to 752 million cubic meters in 2030—and, bear in mind, these figures cover only wood burned as fuel.[17] That is twelve times the average annual harvest of wood in Germany and the equivalent of about 180 million metric tons of oil. To put this in perspective, the EU as a whole used 705 million metric tons of oil in 2019.[18] Wood, therefore, is poised to overtake oil as an environmental pollutant. The problem is not just that wood spews out carbon dioxide when it's burned. The forest floor, stripped of its trees, also releases massive quantities of carbon dioxide as the microorganisms in the warming soil speed up and consume every last scrap of humus. Between the two processes, that's an enormous amount of greenhouse gas.

The destruction of large forest ecosystems with the resulting loss of regional cooling and rainfall probably has such a strong influence on climate that even today wood should be mentioned in the same breath as oil. But that needs to be more closely studied by science. Which brings us back to the beginning of the debate. As long as influential scientists in the forest industry refuse to acknowledge even the most obvious connections, we're not going to make much progress.

Forest agencies are complicit in this denial. That's hardly surprising because these supervisory bodies, which are supposed to stop the overexploitation of forests, are the largest sellers of wood in Germany. This bears repeating: the authorities are, in effect, overseeing themselves. There have been many judicial rulings aimed at curbing the eagerness of state forest agencies to cut and sell as much wood from forests as possible.

Before 1990, we had what was known as the German Timber Promotion Fund. The money was used to promote wood as a raw material to encourage sales and ensure more trees were cut. Anyone who sold wood had to pay a percentage of their income into a communal pot. Contributions were compulsory and the percentage was set by the state. People were being forced to promote the exploitation of forests whether they wanted to or not. In 1990, the Federal Constitutional Court ruled that the levies were unconstitutional, partly because managing public forest lands was not supposed to be about selling wood but about protecting and restoring forests.[19]

And what was the result of this ruling? The law was tweaked a bit, and then it was business as usual once again. It wasn't until 2009, after the Federal Constitutional Court had once again denounced the practice as unconstitutional and banned it, that the compulsory contributions to the German Timber Promotion Fund were discontinued.[20] What was not discontinued was the practice of making timber production in public forests the centerpiece of forest management. We can only hope it won't be another nineteen years before the court hands down a third judgment.

THE STATES ARE NOW having to deal with a completely different damper on their efforts to sell enormous quantities of wood. For years, the Federal Cartel Office has been trying to prohibit the sale of timber through forest agencies, arguing that they are basically a monopoly without any real competition.[21] After many years of trying unsuccessfully to shut down these timber sales and making little progress, the Federal Cartel Office has handed the matter over to a US financial services company. Burford Capital is suing German states on behalf of German sawmills for commission and is demanding compensation of 183 million euros from the state of North Rhine-Westphalia alone—a huge sum for the small forest sector.[22] Burford Capital is representing the Compensation Company for the Sawmill Industry Rhineland-Palatinate, Limited (ASG 3 Ausgleichsgesellschaft für die Sägeindustrie Rheinland-Pfalz GmbH) on the same terms and is suing the state of Rhineland-Palatinate for 121 million euros in damages. The state minister of the environment, Ulrike Höfken, is complaining that the suit is having a devastating effect on forests.[23]

Court cases drag on, and so far the forest industry has managed to use every loophole to maintain the status quo. Meanwhile, climate change marches on and precious time is being lost—time we should all be using to take action. And we can do this. Let's leave the forest and go back to the familiar territory of our dining rooms.

22

What's on
Your Plate?

THE HEADLINES ABOUT climate change are dominated by pollution spewing from pipes. Whether they come from vehicle exhausts, chimneys, or aircraft engines, carbon dioxide–filled emissions have been at the center of the climate debate. Add photographs of melting glaciers in Antarctica and blazing forests in the Amazon, and you have perfectly captured the gruesome apocalypse. The advantage of this approach is that although we all know people everywhere are affected, for the majority of the Earth's population, the problems are playing out on their television screens.

In Germany and other places in the world, however, temperatures are rising and drought is increasing for a completely different reason: we are converting forests into cultivated landscapes. You already know about the cooling effect of forests, as well as the up to 18-degree Fahrenheit (10-degree Celsius) difference between forested and agricultural areas (and an even greater difference in cities). And this difference in temperature begins on your plate.

I've compiled a few numbers to illustrate the problem. And don't worry. If you work through the calculations with me, the results will make you feel optimistic. By the time we're done, I will have shown you one of the most important and at the same time most achievable solutions in our fight against climate change—I promise! To keep things simple—and because science works with the metric system—I will stick with metric measurements for these calculations. The most important take-home message is the relative numbers.

PEOPLE PURSUING PROFIT have reduced Germany's forest cover to 32 percent of the total area of the country, and what forests we have left have, overwhelmingly, been converted to plantations. Even more obvious are changes to the rest of what was once primeval forest: 14.7 percent has been taken over for settlements and roads, and a few more percent are bodies of water, open-pit mines, and waste ground. The lion's share, however—47 percent or 167,000 square kilometers—has been set aside for agriculture.

Of the land set aside for agriculture, 47,000 square kilometers are used for growing crops such as potatoes, wheat, fruit, vegetables—even wine grapes. Then you have fields for biofuels and biogas. Growing these fossil fuel replacements takes up 20,000 square kilometers. Finally, 100,000 square kilometers are farmed to feed the animals we use for meat, eggs, and dairy— almost as much as the total forested area (114,000 square kilometers).[1] Although Germany is self-sufficient when it comes to many plant staples, for animal husbandry, we rely on enormous areas in other countries, where, for example, soybeans and other high-energy crops are grown for animal feed.

I am going into so much detail about land use because these numbers are key when we talk about how eating meat contributes to the greenhouse effect. Many calculations consider only the amount of carbon dioxide the production process contributes directly, while ignoring the effect of converting forests into pastures and agricultural fields. To explain this carbon footprint in a way that's clear and easy to follow, I'd like to walk you through a rough estimate. We're not talking exact numbers here; we're just getting a general idea of how it all works.

First, let's look at how much carbon dioxide is stored as carbon, on average, in the biomass of a forest. In an intact primeval beech forest in central Europe, it's about 1,000 metric tons per hectare.[2] If you convert a forest like that into grazing for cattle, most of the carbon from the felled trees and from the ground escapes into the atmosphere as carbon dioxide and therefore has to be calculated into the carbon costs of beef production.

Now, you could argue that the forest was cut down centuries ago and should, therefore, no longer be part of the calculation. Opinions differ, so to be conservative, let's make the calculation based on the grasslands we have today. These could either be used for cattle or they could be reforested. In the first scenario, the carbon dioxide taken up by grass as it grows every year mostly ends up in bovine stomachs and escapes back into the atmosphere as the cattle digest their food. But if the pasture is planted with trees (or allowed to revert to forest naturally), then most of the greenhouse gases absorbed by the trees will be stored as carbon in wood and humus.

What quantities are we talking about here? Grasslands and forests store astoundingly similar amounts of carbon per hectare per year, namely 6 to 9 metric tons (grass) and 4 to 7 metric tons (forest). To simplify our calculations, let's use 6 metric tons for both. Carbon is converted to carbon dioxide by a factor of 3.67,[3] so if you have 6 metric tons of carbon, that means the plants have removed 22 metric tons of carbon dioxide from the atmosphere. A portion of this is released back into the atmosphere by animals, fungi, and bacteria feeding on grass, humus, dead trees, and so on. However, in the forest, the equivalent of at least 11 metric tons of carbon dioxide is stored in trees as they grow new wood.[4] In total (including bark, leaves, and humus), we can confidently use the figure of 15 metric tons a year. On the flip side, if livestock is kept on this hectare of land, the area can no longer store 15 metric tons because the trees have been replaced by grass that is constantly being grazed by animals. This amount of carbon dioxide must be calculated as a cost of livestock husbandry. Now let's look at the cost per kilogram of meat.

The hectare in question can feed an average of no more than one animal weighing about 500 kilograms. When it's slaughtered, you end up with 53 percent of the carcass as meat, which comes to 265 kilograms. And so, for 265 kilograms of meat, 15 metric tons of carbon dioxide has been released from the area that is now grass and no longer forest. This comes to 57 kilograms of greenhouse gas per kilogram of meat. The overall carbon footprint is much worse because agricultural machinery is used to make hay and the carcass still needs to be transported to the supermarket after it has been processed. Moreover, every day of its short life, the animal expels 200 liters of methane,[5] a greenhouse gas that is twenty-one times as detrimental to the climate as carbon dioxide.

If we want to delve back into the past, we could also add to this calculation the 1,000 metric tons of carbon dioxide that were released into the atmosphere when the original forest was cleared. If we spread this over two hundred years of using the land as a cattle pasture, we can add 5 metric tons or 19 kilograms of carbon dioxide per kilogram to the climate costs of beef. Other carbon costs associated with raising cattle—where the feed comes from and how its processed—contribute more than 20 kilograms of carbon dioxide per kilogram of beef.[6] (The exact amount varies depending on which model you use for your calculations.) This brings us to just under 100 kilograms of carbon dioxide per kilogram of beef.

I'll repeat that this is a rough calculation using maximum values so you can see the kinds of numbers involved in eating meat. The amount of meat eaten in Germany each year comes to 87.8 kilograms per person (or about 60 kilograms by the time it lands on your plate),[7] which amounts to a whopping 8.8 metric tons of carbon dioxide per person per year in Germany—just for meat consumption.

According to the German Environment Agency, the carbon costs for all food consumed in Germany in 2017 came to 1.74 metric tons per person per year.[8] Meat clearly was assigned a far lower carbon cost that did not take into account the loss of forests in Germany. Some sites put the cost of South American beef, which is produced on pastures that were once rain forest, at a whopping 335 kilograms of carbon per kilogram of beef,[9] which is three times the rough estimate that we've just worked through.

Not all people who eat meat eat nothing but beef, which turns out to be particularly bad. Pork and chicken are considered less damaging to the environment, although in most calculations once again deforestation is either not fully accounted for or ignored completely. Without it, the most important element in the equation is missing, the figures become practically meaningless, and the public loses sight of the primary importance of food in our fight against climate change.

STRICTLY SPEAKING, here in Europe it is not so much deforestation as barriers to reforestation that make the current balance sheet for meat production look so bad. That also explains why many media outlets overlook the issue. A former forest that is now a meadow or pasture looks idyllic without any suggestion of climate catastrophe. Smoking chimneys are a call to action; grassy landscapes with butterflies flitting over them are not.

I'd like to try a small thought experiment with you. What would it be like if your consumption of meat was reduced to the once-popular Sunday roast? How many forests could then regrow and what would that do to temperatures in the future?

Let's figure this out using numbers from Germany. One serving of meat generally weighs about 150 grams. If we calculate that out over a year, the amount of meat consumed drops from 60 kilograms to 52×150 grams = 7.8 kilograms of meat per person per year, which means meat consumption would be reduced annually by 52.2 kilograms or 87 percent. We could then free up the corresponding amount of space now used to grow feed and allow that land to once again become forest. Before anyone mentions that a considerable portion of animal feed is imported: that's the beauty of this percentage calculation. If you reduce the need for animal feed by 87 percent, you can reduce the corresponding areas now used to produce this feed either in Germany or anywhere else in the world.

IF WE ATE LESS MEAT, we would need to make space to grow more plants to replace the calories we are losing. No problem: we can use the areas currently being used for biofuels and biogas. Growing biofuels is as damaging to the environment as producing meat, as I learned when I was researching a book about bioenergy back in 2008. A biogas reactor is basically nothing but a huge artificial cow. Grass and maize silage is fermented in vats. Carbon dioxide and methane escape either directly through leaks or indirectly later when the fuel is burned. These fields should be repurposed right away. We could use them to grow organic crops to cover our extra nutritional needs.

This leaves us with our best-case scenario in which meat consumption drops by 87 percent or, to put it another way, in which 87 percent more land can be reforested. If 100,000 square kilometers are being used to grow animal feed, that means we now have 87,000 square kilometers available for new trees. In Germany that amounts to 200,000 additional square kilometers of forest, which is 56 percent of the land in the country.

The transformation of the landscape directly on our doorstep has an additional advantage: if we literally see how reducing our consumption of meat leads to a return of large forests, cooling of the local climate, and increased rainfall, this might fire up the political will to finally put an end to the factory farming of animals.

This is already happening in the Netherlands. The government is promoting an end to factory farming with a program that rewards farmers with up to 1.9 billion euros in compensation over ten years if they rip out their barns and invest in other activities, such as tourism.[10] I would like to see a program like that in Germany. We produce 8.6 million metric tons of meat a year.[11] That's more than 100 kilograms per person and considerably more meat than we consume. Germany produces a lot of cheap meat for export using imported feed, some of which is grown in fields cleared from tropical rain forests. I think the Netherlands' approach of using financial compensation to encourage owners of factory farms to gradually shut down this gruesome business is a reasonable compromise. The money is well spent because it considerably reduces environmental costs and thus the costs for all of us in the future. Think of our groundwater, our most important food resource, the quality of which is currently deteriorating as it is being flooded with liquid manure.

A forest could be a much more relaxing source of income for farmers than the ever-tightening market for cheap meat. If farmers could earn 1,000 euros per hectare per year simply by allowing their forest to grow, then not only would their bank accounts fill themselves, but their public image would also be much improved. Their new role comes with a job description at no extra charge: climate guardian.

ASSUMING WE ALLOW large forests to return, what happens to the animals that currently live in open grasslands? Many times during tours I've led through the forests of the Eifel, participants have angrily protested when I suggest more pastureland should be reforested. Open fields, they tell me, are important for many herbs and grasses, for insects and amphibians, all of which are being crowded out in our managed landscapes. As noble as this argument in defense of nature is, the assumptions on which it rests are incorrect.

Native species in Germany that live in open spaces and frequent forest clearings—many insects, for example—usually cannot survive in livestock pastures. They require grassy meadows. And there is a big difference between the two. Grassy meadows are grazed—or better yet, lightly nibbled— by large, wild herbivores that roam over large areas. These meadows once occurred naturally in the floodplain forests that used to stretch for miles on either side of large rivers in central Europe. As late as the middle of the twentieth century, the rivers froze regularly and that was an opportunity for natural meadows to form. Drift ice in spring created open areas between the trees where grasses, herbs, and bushes could grow. In these semi-open spaces dotted with trees, wisents, aurochs, and wild horses once grazed, creating a grassy landscape filled with a wide array of plant species that were vital to the survival of thousands of species of insects.

Apart from a few pitiful remnants, floodplain forests have disappeared, and, with rare exceptions, the floods are also long gone—to say nothing of the ice. There's one main reason the ebb and flow of these valuable forests has practically disappeared: people have displaced the wild animals that once grazed in these river valleys. The most productive agricultural fields are to be found here because with every flood the river deposited nutrient-rich mud. This is where we built our settlements and towns. This is where our civilization spread. Floods are a threat to cultivated spaces and therefore water is restrained by tall dikes and dams. You can still find pathetic remnants of floods in the form of retention ponds and drainage ditches, although they are used for temporary water storage and are not suitable for the establishment of true floodplain forests.

AND SO, IF WE WISH to do something in Germany for animals that live in grassy spaces, then it should be done in the places they have always lived. We urgently need another national park in the Rhine Valley or along the Elbe. The national park in the lower Oder Valley is a small start, but at only 40 square miles (100 square kilometers), it really is tiny. And even here, only parts of the park are left to nature. Under pressure from a combined lobby of farmers and anglers, only 50.1 percent of the area was reserved as a natural space; the rest of the park is still open for business. The figure of 50.1 percent was purposefully chosen because if an area is to be a national park, more than half of it must be strictly protected—at 0.1 percent, the margin is razor-thin, but the park can still be called a national park.[12]

We still don't have a true, large riverine wilderness that offers the trees of floodplain forests and their wildlife communities refuge. Instead, we dot the central uplands with pastures, which are then leased out for flocks of sheep to graze. These hills, however, used to be covered with ancient beech forests, and wild cattle and horses were never present in large numbers. And yet this is where a project to protect a species was established.

Richard zu Sayn-Wittgenstein-Berleburg initiated a project to reintroduce wisent.[13] The forest where the wisents were to live was not in a floodplain but in the Rothaar Mountains, a range of hills in Sauerland. The prince set aside about 15.5 square miles (40 square kilometers) in the uplands where the wild cattle were to roam. But 15.5 square miles, although it sounds like a lot, is not nearly enough for animals that weigh almost a ton. And so what was bound to happen happened: the animals did not remain in their designated area but happily wandered through pastures, agricultural fields, and forests. On their travels, they gnawed on tree bark, which reduced the trees' economic value. No wonder forest owners complained and demanded compensation and the removal of the wisents. Now the herd is to be culled and fenced in, and the animals will be living in what amounts to a wild animal park. This is not what nature looks like.[14] Large herbivores like the wisent are another reason we desperately need at least one large national park in a floodplain forest.

IT WILL CERTAINLY be a while before politicians move in the direction of less meat, more forest, and additional national parks. At least with meat, everyone can make a start for themselves. And if you were wondering, I completely stopped eating meat in 2019. Besides the fact that my wife and I do not like to see animals suffer, what convinced us to change our diets was our concern for the planet. And there are other things you can do, right now, on your own doorstep. Read on.

PART III

Forests of the Future

23

Every Tree Counts

A SINGLE TREE? What good is that? I get this question increasingly often. On a global scale, planting a single seedling is certainly less than a drop in the proverbial bucket in the fight against climate change. Apart from that, I'm convinced that in many places forests can return all by themselves. On a local level, however, planting a tree is a very different proposition—and here I mean really local, as in right in front of your house. Planted there, a single tree can have a beneficial and measurable effect on weather conditions. This is something you can test for yourself.

If you park your car under a tree in winter, the windows don't freeze as quickly. This is due to the way trees moderate temperature extremes: under the crown of a tree, just like under a roof, it doesn't get as cold. In summer, the opposite happens, and the tree will cool things down. Not just because it casts shade but also because it transpires water, which lowers the ambient temperature. Here again, you can do a little personal experiment. Open up a sun umbrella on a hot summer's day and sit underneath it. You'll find it's still hot under the umbrella but not as hot as it would be without it. Now sit under a tree and feel the difference. On average, especially large old deciduous trees lower the temperature by about

3.6 degrees Fahrenheit (2 degrees Celsius). This is not surprising when you remember that an old beech releases up to 130 gallons (500 liters) of water from its leaves, water that uses heat energy from the surrounding air as it evaporates. Our bodies achieve the same result when we sweat to cool ourselves down.

If the tree is close to your home, the enormous amount of water evaporated often ends up as condensation on the walls. Grayish-green algae may begin to grow in the shadow of the crown, indicating that the air there is especially humid. I've experienced something similar at our forest lodge. As the name suggests, the house is surrounded by trees. One big, sturdy birch stands out. It's the largest birch I've ever seen. It grows 25 feet (8 meters) from my office window, and its hollow trunk offers birds a safe place to build their nests. The weather station at the lodge regularly shows a temperature difference of 3.6 degrees Fahrenheit (2 degrees Celsius) compared with the temperature at the forest academy, which lies on the next ridge of hills in the neighboring village of Wershofen. The temperatures differ because the trees at the forest academy are still small—the buildings and the grounds were completed at the end of 2019. On hot days, it's 3.6 degrees Fahrenheit (2 degrees Celsius) cooler at the forest lodge, and on cold days, it's 3.6 degrees Fahrenheit (2 degrees Celsius) warmer. It is also more humid at the lodge. The differences, which we notice year-round, can mostly be attributed to the old birch and the other old trees in the yard.

A SINGLE TREE, then, certainly can have a direct influence on the microclimate at your house. I find this worth mentioning because every tree growing in a front and backyard contradicts the depressing idea that individuals can never change anything.

Which tree is the best one for greening up backyards or streets? It should always be a native, because here, as in the forest, the food chain depends on trees and the trees, at least partly, depend on the food chain (remember the holobiont?). It's a good idea, therefore, to check out which trees are growing naturally in forests near you. In Germany, you'll find oaks, beeches, field maples, wild service trees, or quaking aspens, all of which are bravely resettling clear-cuts all over the country. If you want trees that serve more than one function, fruit trees are a good choice. Children especially enjoy growing up with trees around them. For the rest of their lives, they will carry with them the instinctive feeling that these gigantic beings are important to us.

There were touching moments on display in many cities over the past few dry summers. City dwellers were concerned about their street trees and began to water them. And they did not undertake this activity alone. Caring people organized community watering brigades by street, drawing up watering schedules for their treasured trees. It was a hopeful sign, and a sign that oaks, plane trees, and maples are increasingly valued and no longer viewed as simply green decorations. The call to action was often empathy for the thirsty giants. Watering can by watering can, life-saving water was poured over the dry soil around the trees. But was this enough?

Our forest academy fielded many questions about whether the assistance was really going to help. To answer them, we should check to see what nature has to say. A shower that delivers up to half an inch (up to 1 centimeter or up to 10 liters per square meter) of rain barely penetrates the surface when the soil is dry. When only half an inch (1 centimeter) of rain falls, it's clear that the moisture can seep no more than an inch or so (a couple of centimeters) into the ground. But delivering even such a small amount of water can be overwhelming for watering brigades. The roots are not confined to the small area directly under the decorative metal grate around the tree. They usually spread over an area twice as wide as the crown. The crown of a mature street tree can easily be as wide as 32 feet (10 meters), which means the roots spread for 64 feet (20 meters). If you do the calculations, you get a root area of 3,380 square feet (314 square meters). To pour half an inch of water over an area this size (10 liters per square meter), you're going to need about 800 gallons (approximately 3 cubic meters) of water, which would be too much for any watering brigade to carry.

Even if they could deliver that amount of water, the brigades would never be able to reach all the roots. It is the nature of a city that large areas are covered by surfaces such as concrete and asphalt, which water cannot penetrate. Often, all city planners allow a tree are its meager planting spot and the circle of soil immediately around its trunk. Does it make sense to occasionally pour a can of water over this spot at least? Absolutely! Imagine you are crossing a desert and about to die of thirst. You need many gallons of water, but you have none left. Wouldn't it be lovely if a helpful person were to offer you at least a mouthful? Also, watering as a group creates a sense of connection that can be passed on to others. It helps promote empathy in our society and in the long term that will help us get more forests.

HERE'S A COMPLETELY DIFFERENT WAY to help trees return to places they once called home and allow them to cool the surroundings once again: agroforestry. It sounds technical but in practice it is very simple. Trees and crops are allowed to grow sometimes closer together and other times farther apart. This system has many advantages both for the crops and for nature.

Let's first look at the crops. The majority don't thrive in the shade of old trees, as their ancestors mostly grew in open grasslands and needed full sunlight. There's less sunlight the closer you get to trees but a lot more a little farther away in treeless fields and pastures. Out in the open, nearby trees provide a windbreak for crops. Sheltered from breezes that dry out the soil in summer, the topsoil remains moister. And moisture, as we have learned the hard way over the past few summers, is the key to agricultural production. In periods of drought, even the shade from the trees can be helpful. In the dry summers of 2018 to 2020, shaded areas were the only places where grass in the pastures remained lush and green, and cattle could find sanctuary under the trees and cool down at least a little.

Another benefit trees offer cultivated crops is what is known as hydraulic lift, where trees function as a kind of water pump for other plants. The roots of grains, potatoes, and other crops tend to remain in the upper layers of soil. Unfortunately, this is the first layer to dry out, as you probably know from your own garden beds in summer. Even when the upper layers turn hard and grainy, you often find moister soil just 2 to 4 inches (5 to 10 centimeters) down. Depending on soil conditions, this layer of moisture might extend down many feet, but the roots of our field crops and grasses don't reach down this far.

Trees, however, have no difficulty extending their roots into this area. They can pump up enough water from deep down to take care of themselves. Mature beeches and oaks tip the scales at more than 20 tons—a biomass that needs to be supplied with many hundreds of gallons of water a day in summer. And so their roots, aided by cooperative fungi, draw an enormous amount of water from the ground.

During the day, this water finds its way to the end users—the trees' leaves—where sunlight combines it with carbon dioxide to make sugar. A large amount of water, however, escapes through the minuscule, mouthlike openings on the underside of the leaves into the forest air, where it cools the whole ecosystem. At night, however, the leaves shut down and all activity stops. Aboveground, nothing is happening—except for one thing. The giants' trunks increase slightly in girth because the leaves are no longer accepting water.[1] The trees' tissues fill with moisture until, finally, there's no more room—a trunk, which is made of wood, can only expand so far. In many cases, however, the roots continue to pump water from the ground. When the trunks are full, where does it end up? Todd E. Dawson from Cornell University in Ithaca, New York, studied this phenomenon with native sugar maples. He discovered that the ground up to 16 feet (5 meters) out from the trunk was distinctly moister at night.

Hydraulic lift benefits the tree because humus on the forest floor stores a particularly large number of nutrients. Humus is formed as plant debris rots. Most of the debris falls from above and lands on the ground, where earthworms and other organisms break it down. During the process of decomposition, many nutrients are released, which plants can only take up after they have been dissolved in water—and the trees provide the water they need. What a practical arrangement!

Researchers have found hydraulic lift in European deciduous forests as well. They investigated a young beech-and-oak forest. To simulate extreme drought, they covered their test forest with a roof so the ground dried out. They attached probes to the trunks to siphon off water being pumped up. They then piped chemically marked water 30 inches (75 centimeters) below the surface and observed how the trees reacted. It didn't take long for the researchers to find marked water in the trunks of the deep-rooted oaks. But they found none in the shallower-rooted beeches.

While the middle soil layers remained dry, the marked water was still turning up in the upper layer of soil six days later. The water could not have wicked up through the soil by capillary action—if it had, the soil would have been moist all the way up. Although the researchers did not find any evidence of water being exchanged between tree species, in their opinion, oaks might make an important contribution to keeping forests going in times of drought. (Because the measuring apparatus was expensive, the French team studied only four trees and the researchers were unable to determine conclusively whether the beeches benefited.) They believe more than just the trees benefit from the moisture in the upper layers of soil. A multitude of other species including other plants, fungi, bacteria, and soil-dwelling organisms do as well—and together they keep the ecosystem, including the beeches, in good condition.[2]

A small sidenote here. Natural beech forests consist principally but not exclusively of beeches. Many other species of trees grow in these forests, especially oaks. Even if the two species don't necessarily work together directly, each might be stronger when they grow together, especially in these times of climate change.

LET'S GET BACK TO agricultural spaces. Trees encounter huge problems here. We've already talked about how difficult it is for roots to grow in compacted soil where oxygen is scarce. Unfortunately, most agricultural soils are like this—after all, who works with horses these days? Every square yard is driven over by heavy tractors hundreds of times and densely packed. The time it takes for soils to recover from this damage—if they ever do—can be measured in millennia. Frost (when water expands and loosens the soil) and digging and burrowing by animals large and small can open up the soil a bit—but only in the upper layers. But here, too, research results from Todd Dawson give us some cause for hope. The strong roots of the trees he studied penetrated the compacted layer, and at night, they pumped water out of the soil beneath it to the upper layers, where it was available to the shallower roots growing in looser soil.[3]

Nothing in nature happens by chance. Pumping water to the surface is an energy-intensive activity. The fact trees continue to do this at night has many advantages, especially in dry summers. Water from deep underground is delivered to the many fine feeder roots growing just under the surface. Each morning, the trees can basically have their breakfast right away and jump right into photosynthesizing. To do this, they need not only the water but also the nutrients dissolved in it, which are absorbed by the fine roots as the trees drink.

It's a wise move on the trees' part to irrigate the ground at night, as you can see in your own garden. If you have a garden, you probably also know that the best time to water is in the evening when the sun is no longer shining, the air is cooler, and water doesn't immediately evaporate. This allows moisture to gradually seep into the ground, where it is ready and waiting for the plants the next morning. When trees water themselves, they adopt a similar strategy. This strategy also helps them conserve energy. If trees were to water the ground while also providing water for photosynthesis and transpiration, they would have to pump much harder during the day than at night. They have organized it so that the pump hums along at the same rate day and night, serving different purposes at different times.

If we use trees to help us in agriculture, we get a bit of nature back into the bargain. Strips planted with trees provide shelter and food for birds and many other animals. The trees restore a piece of the much-abused fields' wild souls—and that alone makes planting trees worthwhile.

If the advantages of trees are so obvious, and it's clear conventional practices aren't working for forests, then why is it taking so long for things to change? Could it be that too often we wait for the last hard-liners to get on board before we take action?

24

Does Everyone Have to Be On Board?

IN FALL 2020, I was part of a group of environmentalists discussing a model we could use for best forest management practices. With the endless cycle of self-destructing plantations, the rush to remove damaged timber, and the subsequent replanting of trees, we wanted demonstration and test plots to showcase alternative methods of forest management. We also wanted to draw up a list of ecological management practices.

During our discussions, someone asked if it was okay, during the transition from traditional to ecological management practices, to allow the use of feller bunchers—that's to say the use of heavy machines that can fell and process trees on the spot. The fact that this question even came up really annoyed me. The conciliatory attitude some environmental groups have taken vis-à-vis traditional forestry has done nothing to stop brutal harvesting techniques over the decades. Quite the opposite. The heaviest machines began their triumphant march in 1990, and the size of clear-cuts was only slightly reduced for a while and are now larger than they have been in decades.

Given this history, I think the time to make allowances for forest companies that want to improve their public image but are not willing to do anything meaningful to stop destroying the soil is long gone. During the discussion, someone also suggested we needed everyone on board, an approach that clearly hasn't gotten us anywhere when it comes to forests.

TO WAIT FOR EVERYONE is to match the pace of progress to the slowest among us. In the past few decades of environmental politics, we've experienced what waiting to have the last doubter on board means. Despite technological innovations, the global output of carbon dioxide has continued to climb, and even the coronavirus pandemic has not brought any major changes in direction.

NGOs have also had minimal success when it comes to forests. Despite unending rounds of dialogue and some violent protests, forestry practices have not improved. As I just mentioned, clear-cuts today are the largest they've been in decades, even though it flies in the face of the policies of every state in Germany. Certainly, a few patches of forest are being managed in an exemplary fashion—the forest near the city of Lübeck is a case in point. But these patches are few and far between. Meanwhile, forest management practices are becoming increasingly heavy-handed as gigantic machines advance through the trees and helicopters douse enormous areas of forest with poison.

The crucial point here is that no one is talking about where forestry has gone wrong—not to assign blame but to establish that traditional methods are not working. But this level of insight is missing, and all the blame falls on climate change. And because anyone out for a walk can see that current management practices are failing on a large scale, our forest guardians in green are spreading disinformation about why all this was inevitable.

Foresters swear that the dead and dying conifer forests are their predecessors' fault. After the Second World War, foresters were tasked with providing timber for the reconstruction of Germany, which led to the planting of enormous monocultures of spruce and pine. Apart from the fact that conifers continue to be planted to this day, this practice has a much longer history. The American forester and conservationist Aldo Leopold visited Germany, land of his forebears, in the 1930s. He noticed then that much of the famed German forest consisted of unnatural conifer plantations where wild game was kept so people could hunt. He called this disaster "the German problem." And the problem persists to this day.

According to the latest federal forest inventory, which took place in 2012, almost none of the much-vaunted conversion of plantations to more natural forests has happened. Our most important tree species, beech and oak, come in at only 15 and 10 percent, respectively. If forest conversion really had been happening full-throttle for decades, we would expect to find an especially large number of these deciduous trees in the youngest stands—those where the trees are less than twenty years old. But far from it—their share of the inventory in this age cohort was only 12 and 6 percent, respectively.[1] Since Aldo Leopold's time, forestry has at best been marching on the spot.

HOW DO WE UNTIE this Gordian knot? We should cut right through it or, as the head of the city forestry office in Lübeck, Knut Sturm, said on the radio, we should "take the forests away from the foresters!" It won't come to that, of course, but we urgently need a completely different approach to training the guardians of our green lungs. It will be a long and rocky road, for sure, but despite the difficulty, a few of us will soon be embarking on it—more on that in the next chapter. For many forests, the winds of change will be blowing through too late. After all the old trees have been felled, it takes from decades to centuries for a forest to regenerate. We have run out of time, and therefore we need to use another democratic instrument to protect trees: legal action.

Two conservation organizations, Green League Saxony and NUKLA (a local NGO), have shown how effective legal action can be to aid forests. They took the city of Leipzig to court. The trigger for their action was the felling of trees in a floodplain forest, one of the largest that still exists in central Europe. It stretches for 10 square miles (25 square kilometers) along small rivers, catchment ponds, and channels. Here, too, as you can guess, foresters and city experts tried to help the forest by allowing a lot of logging. Because the Leipzig floodplain forest is a European protected area, this should not happen without an environmental impact assessment, and this was the grounds for the suit brought by the two conservation organizations. On June 9, 2020, the Higher Administrative Court of Bautzen handed down a groundbreaking judgment. The city had to put a stop to logging immediately and undertake a thorough review of its policies. These needed to follow the regulations governing the protected area and be discussed with both conservation organizations.[2]

The situation in Leipzig takes me back once again to the Holy Halls. Around the small preserve containing the oldest beech trees in Germany stretch forests that also enjoy the protection of European law. The law states that their condition should not be allowed to deteriorate. But that is of little interest to the local forestry department. It has allowed so many old beeches in the area to be logged that most of the area is now covered in brush. Unfortunately, this has dire consequences for the Holy Halls. At 165 acres (67 hectares), the area is much too small to cool itself and retain moisture in hot summers. That's why it needs the large buffer of trees around the preserve, and this buffer zone is now severely damaged.

Pierre Ibisch from Eberswalde University confronted the issue with the help of a lawyer. The local forest department ignored the lawyer's letter, and so in December 2020 we used my social media channels to publicize what was happening. Two television channels reacted, and daily newspapers covered the story. Then Mecklenburg-Vorpommern's environment minister, Till Backhaus, got involved. He wanted to avoid negative fallout that might impact tourism in the area and agreed to an online conference call. After our conversation local logging was stopped and a working group was formed to discuss expanding the protected area.

For me, this was a good example that individuals are not powerless. Broadcasting the issue on my social media accounts was effective only because the reports caused such an uproar in the online community (in other words, received so many likes)—every click counts.

With their backs to the wall, the foresters pulled out one last argument, one guaranteed to stir up emotions and shut down logical thinking: timber means jobs. I hear this argument everywhere I go. Whether I'm in Canada, Poland, Sweden, or Germany, this mantra is used to justify the most devastating clear-cuts. If you follow German politics, you might be familiar with this little game in connection with the transition from coal. Fears were fanned, which led to protests in coal-producing areas because a way of life was in danger of being lost. When emotions run high, it's much more difficult to explain that if we continue along our current path, we will all face a far greater loss. Only after payments of billions of euros had been made to the coal industry could peace be restored and a time frame (albeit one way too far into the future) be set for shutting down the industry. All this sounds like a blueprint for other sectors that harm the climate—for instance, forestry.

Forestry is only a small sector in the German economy, and it doesn't carry as much weight as the large energy producers. However, it has a much greater influence on regional weather than any other sector—just think of the cooling effect of forests and the amount of rain they produce. Less important financially and yet with many more negative effects—clearly, we need to reach a political consensus quickly on how to fix forestry. Desperate times call for desperate measures, and those active in state forestry are now adopting the defensive strategy of a toad about to be eaten by a bird: they draw themselves up to look taller and pump themselves full of air. The vehicle they are using to achieve this threatening posture is called the Forestry and Wood Cluster Initiative.

The cluster is an imaginary construct in which all sectors of the industry are combined. And because, on its own, the forestry sector is much too small, everything related to trees is thrown in. Forest workers, foresters, and sawmill staff seem to be a logical grouping. That comes to about 110,000 employees in total, which is tiny when compared with the labor market as a whole. To increase the cluster's political clout, huge sectors such as furniture manufacturers, paper manufacturers, and the entire print industry have been included. A sidenote here: the sectors were not asked if they wanted to join. When I'm giving print media interviews, I like to ask about this—no one I've talked to has had any idea they are a part of this initiative. After these unsuspecting economic giants have been included, the number of people employed increases by a factor of ten to 1.1 million—that's more like it![3] Now the cluster has achieved a critical political mass and can roll out a compelling argument against sparing the forests: every tree left standing costs jobs.

David Suzuki, Canada's best-known environmentalist, told me loggers in that country are more direct. He was visiting Vancouver Island to film a logging camp. Three huge guys suddenly appeared out of the forest and tried to drive off the camera crew. Events then took an unexpected turn, and the two groups fell into conversation. Suzuki told the loggers, "No conservationist is against logging. We only want to make sure our children and grandchildren still have large trees to cut." One of the loggers told him, "My children will not be loggers. By then, there won't be any trees left."[4]

My personal answer to the question posed in the title of this chapter is that no, we don't have to get everyone on board. If we wait for the last hard-liners, we will dilute the process until reform becomes impossible. The hard-liners have had decades to prove that they can manage the forests entrusted to them by the public in a responsible fashion. Unfortunately, the condition of forests today shows they have not done that. Those who have been so wrong for so long now have two options: either admit to their mistakes and change their behavior or accept the consequences and step aside to allow others to guide the forests with a gentler hand on their journey to regeneration.

We no longer have the luxury of waiting decades to see if those now actively involved in forest management will succeed in turning things around. The forest needs a breath of fresh air. That will happen only when we completely change the way forests are managed. The breeze is picking up!

25

A Breath of Fresh Air

IT'S HIGH TIME to change how forestry works. And what better way to do this than by renewing the system from within? It's difficult to teach old dogs like me new tricks, so why not start by teaching young people differently from the start? Even if universities say things have changed, until now, forestry students in Germany have had no choice but to study conventional forestry. The whole training, course of study, and work experience or internship with a state forest agency mainly prepares students for a career in government forestry rather than offering them a wider view of forest management.

State forest agencies are heavily involved in planning the curriculum, primarily through the Federal/State Forestry Working Group. This body is made up of leaders from federal and state forest agencies. They regularly agree on joint approaches to issues that relate to more than one region. In the case of the study of forestry, the working group draws up a catalog of requirements for future graduates.

Public forest agencies dominate not only the timber market but also the labor market in this sector—and the pressure they bring to bear is more than subtle. The extent to which they shape the conversation is clear when you consider the vocabulary they use. Let's consider a few words that make it clear the forest is viewed predominantly as a factory that produces raw material. There is no talk of planting conifers or deciduous trees; instead, the process is described as planting softwood or hardwood. Try that for yourself. You can't plant wood. If you stick boards in the ground, they certainly don't grow. It's like a pig farmer saying he puts pork chops in his pens.

And later, once the seedlings have matured and grown into substantial trees, forests are not described as ecosystems. One of the most important indicators of how well a forest is doing is wood per hectare, which means the amount of wood in cubic meters that is available in the form of living trees. This makes the forest nothing more than a huge warehouse managed by foresters. They check to see if there is enough wood in stock, increase it by planting, and decide when trees are ready to be harvested.

Old trees are described as "harvest-ready," like strawberries that can and should be picked. Unlike red strawberries, trees identified as harvest-ready have lived for significantly less than one-third of their natural life expectancy, so they're more like green fruit. The harvest age is regulated, and it varies depending on what the timber market wants. Thick old beeches and oaks, wonders of nature, are assessed, and when they reach a certain diameter, they are singled out for "end use," which means death with all the grim connotations of that word.

The language that relieves foresters of guilt is employed not only while they are studying. Many foresters go on to use it to defend logging operations in old forests. It's not about sourcing timber, which legally cannot be the reason for the intervention. Rather, foresters say they're helping the poor little beeches that cannot grow properly in the shade of the mother trees. And so it follows that this sourcing of raw materials with massive collateral damage is called "forest restoration." Saying you're restoring forests sounds better than saying you're destroying ancient root networks. It's somewhat ironic that when foresters harvest trees, they claim they're taking good care of forests. It's a bit like butchers telling you they're taking good care of animals.

BEING TOLD THROUGHOUT your studies that the forest is a timber-producing machine dulls your capacity to be amazed by nature. You no longer think it's a bad thing when large harvesters crush the soil and everything in it. It's no big deal when enormous quantities of biomass are removed. After studying forestry, students desperately lack in-depth knowledge of endangered species, something my Swedish friend Sebastian Kirppu sees all the time. He often brings rare species—certain lichens, for instance—to the attention of his coworkers, which has led to many forests being protected despite resistance from the forest industry, making Kirppu one of the most hated environmentalists in Sweden.

At least as unfortunate is that forest owners don't have access to a second opinion when they have questions. Most freelance forest consultants have taken the course and then gone through the mill of state forest agencies, and they regurgitate the messages they have learned almost word for word. I experienced that myself in our forest in Wershofen. In 2018, when we needed to evaluate our forest, as required by German law, we decided to hire a freelance consultant.

State foresters were still focused on same-age forests, that is to say, single-species, plantation-style forests. We wanted to avoid that in Wershofen, and so, on the advice of Wohlleben's Forest Academy, the community decided to consult an independent expert. His conclusions, however, turned out to be a huge disappointment. In a memorable town council meeting, he issued the dire warning that under the influence of the forest academy, the Wershofen forest was gradually turning into a deciduous forest and the plantations of conifers were becoming less dominant. He recommended planting more spruce and Douglas firs so timber companies would not completely lose interest in our forest. He also urgently recommended more logging in our old beech forests. Bear in mind, this meeting took place in May 2018, right before the first of three record-breaking dry years when spruce plantations all over Germany began to die, signaling the end of this species for the forest industry. Obviously, the town council did not follow his advice.

FOR YEARS, PROGRESSIVE FORESTERS have been talking about the need for a new course focusing on environmentally friendly forest management. A breath of fresh air was needed if the profession was finally going to offer students alternatives in the forest job market. This idea had been floating around the forest academy for a while as well, but it hadn't been a top priority while we were getting up and running.

The tipping point came about somewhat by accident. In summer 2020, a team from GEO, which publishes my magazine *Wohllebens Welt* ("Wohlleben's World"), came for lunch. The team was led by editors in chief Jens Schröder and Markus Wolff. They wanted to see our new buildings and talk about the status of the magazine and how the general downturn in the periodical market might affect our work together in the future. It turned out the magazine was bucking the trend and holding its own (hurray!), and the promise was made that we would continue to produce issues in 2021. Our conversation took place on the terrace of the local hotel, where we were not only in compliance with Covid protocols but were also able to enjoy a view of the Aremberg, one of the tallest mountains in the Eifel. The top of this dormant volcano is covered with old beech forests. While we were drinking coffee after lunch and admiring the view, Schröder asked me what dreams I had for the future.

Truth be told, I can't remember my answer, but weeks after the meeting, he emailed me to say he would like to follow up on my dream for my own course in forestry. He suggested we should get Pierre Ibisch on board. Together, we could look for sponsors and a university and get things underway. And then it hit me: my dream was within reach.

Anyone who knows me knows I've always been a fan of moving quickly, and I've made some pretty off-the-wall ideas happen in short order when they mean real progress. At the end of the 1990s, I organized survival weekends in the woods and used the income to save the old beech forests in my community. The trees were due to fall victim to chain saws until I convinced the mayor we could make up for the lost income some other way. Even though the tourism office declined to participate and the forest agency I worked for was somewhat irritated and tolerated my plan at best, "Survival in the Eifel" was a resounding success. It turns out that a forester who leads guests through the woods for days while living off roots and insect larvae is worth a few mentions on television, and no one complained about the advertising and income for the community.

A forestry course of my own, naturally, was a much bigger undertaking, one that also offered many more opportunities.

Let's start with the opportunities. The biggest one is simply that there will be a course like this in the first place. If we were to call it "environmentally friendly forest management," then what would that make all the other forestry courses? In the public consciousness, they would be relegated to the same conservative backwater as conventional agriculture.

To my surprise, we soon found generous donors to fund the positions we needed—a coordinator and two endowed professorships—which meant that there would be almost no cost to whatever university we chose.

What's the best home for a course like this? It's a perfect fit for a university that includes innovation and ecology in its mission statement, and we settled on the Eberswalde University for Sustainable Development. It is one of the smallest universities in Germany, but it has a history of revolutionary ideas about forestry. No sooner said than done. Our first conversation took place in December 2020. It was a bit like poking a stick into a hornets' nest.

Contrary to the wishes of the university dean and president, the department in question refused to even discuss the idea and all kinds of confidential information was leaked. This was followed by ugly statements, both internally and publicly, and news of our project spread quickly to other forestry departments and beyond. Schröder made sure there was thoughtful coverage in the media: a wide-ranging community discussion about forests and their use was, after all, part of what we wanted to achieve.

At the core of the ugly defensive reaction from forestry departments was fear about how the public would view what they offered. In a joint statement made by colleges and universities that already offered courses in forestry, the signatories suggested that the project should be abandoned.

After all, environmentally friendly practices were already a central part of existing programs. If that really were the case, then the traditional schools of forestry could look on unconcerned as lack of demand caused our new course to fizzle out. Incidentally, the declaration states it was signed by entire departments and universities, giving the impression of unanimity in the opposition to our project.[1] Fortunately, that isn't completely true, as we learned when we received encouraging feedback from a number of these universities.

Another issue emerged that might worry young students. What if they backed the wrong horse? What would happen when those trained in the traditional way suddenly found themselves competing with students graduating from our course? Fear often overestimates the perceived danger. We're talking about twenty to thirty places a year in the program we are planning.

We don't want confrontation, but we do want to address these fears. Forestry as traditionally practiced has reached the end of a long and destructive road, as anyone can tell just by looking at all the clear-cuts. The forest is the best testament to traditional courses in forestry—either these are still too focused on managing plantations or no one is putting what they have learned into practice in a way that improves forest health. If the trees were in charge, they would give the universities a failing grade. It is time to completely reform forestry studies so young people are adequately prepared to fundamentally change how forests are viewed and managed. Luckily, it's not too late to do that. Despite all the damage inflicted on them, forest ecosystems are still adaptable and strong.

26

The Forest Will Return

I PURPOSELY SAVED the good news for last. If we allow them to, forests will return, at least in places where forests still grow, even if the trees right now are suffering.

Forests have always needed to regenerate because catastrophic events have been their constant companions for hundreds of thousands of years. Depending on the region, catastrophes pop up rarely or often (measured by the life of a tree). Deciduous forests in eastern North America have always been particularly battered, as they still are today, because the main mountain ranges run north to south; there are no mountains ranges oriented from west to east like the European Alps. Storms in which warm air from the south and cold air from the north meet can be especially turbulent, and even forests of beeches, oaks, and maples often don't live more than a hundred years before a tornado blows them over.

The situation in Europe is different. Here, deciduous trees in ancient forests have the luxury of living for five hundred years or more before they are toppled. Storm damage that flattens acres of trees does happen, but it is rare. And yet the community of trees grows back—if it is left alone and allowed to do so in peace.

Many people, however, still resist this free (self-)help. The desperate way some plantation owners fight to hold on to their beloved softwood production is almost moving—when I think of the plight of the forest. For years I've been watching the decline of a spruce forest close to the forest I manage. It exemplifies the whole situation forestry finds itself in, while simultaneously pointing to new opportunities.

In the summer of 2018, bark beetles descended on a small corner at the edge of the plantation. The crowns of the dying spruce stood out for miles as the trees fought to the death and their needles faded from green to rust. As you now know, the best thing would have been to leave the trees standing— at least the ones that were completely dead—because bark beetles are not interested in dead trees. But the owner cut the trees to clean up the forest. The next winter, a medium-sized storm hit the forest. The hole at the edge afforded a good point of attack. The battle-hardened trees along the perimeter, which had acted as a kind of bulwark to dampen the force of the wind, had disappeared, and the swaying crowns of the next row of spruce were left without any protection from that side. Hundreds more trees toppled over. In the spring, the owner cleaned up again and removed the trees so the ground was back to being neat and tidy. But an inexorable process had been set in motion, and a year later most of the remaining spruce also fell to the ground in a late-winter storm.

Meanwhile, the timber market in Germany had collapsed—because the same thing had happened to many forest owners. Moreover, bark beetles attacked the miserable remains of the plantation, which spurred the forest owner to act even more quickly. Finally, the forest was clear-cut, leaving only the roots of downed trees as a mute testament to the storms. Now would have been the time to do things differently, but no, it had to be spruce once again, which were now replanted along with Douglas firs. "How can anyone be so blind?" was the question that ran through my mind.

By spring 2020, deciduous trees and many herbs were sprouting up between the dead-straight rows of conifers stretching up the mountain slopes. It was as if nature was shyly drawing attention to itself and offering free help. No matter, the man fought for every conifer. In late spring, he freed the spruce seedlings from the lushly sprouting vegetation, carefully snipping away everything green up and down the rows. Nature's answer was not long coming. First, the new growth on many of the conifer seedlings froze in a late cold snap in the middle of May. The plants around the little trees would have been able to insulate them from the cold, but they were no longer there. Then it got hotter and drier. The seedlings were sorely in need of a bit of shade. Many of them died the year they were planted—unlike the free gifts from nature, thousands of aspens, birches, willows, and beeches.

The drama isn't over yet, and there's still hope because nature has time on its side. Even if the owner once again cultivates a plantation of conifers, denying the inevitable failure of conifers as a crop, nature will continue to offer assistance. Every year, new deciduous trees sprout and grow happily despite climate change and dry summers, demonstrating that they are a free and better alternative. Even though the clearcut should put me in a bad mood, I can't help smiling every time I pass by.

The man's actions are understandable when you consider that most forest owners follow the recommendations from traditional forestry. Even in 2020, the chair of the Scientific Advisory Board for Forest Policy at the federal Ministry of Food and Agriculture, Jürgen Bauhus, clearly did not believe that after hundreds of millions of years nature could regenerate forests on its own. In an interview with the *Stuttgarter Zeitung*, he uttered a couple of sentences that combined the situation forestry finds itself in with the arrogance of its experts: "It [the scientific advisory board] prepares reports based on scientifically proven findings. It cannot allow itself to be in the position of providing political advice based on evidence-free narratives such as the power of nature to heal itself."[1] Think about those words for a moment. The most important group currently advising politicians about forests is telling them nature no longer has what it takes.

It that were true, forests would be lost without people to help them. How can the forests in the endless Siberian taiga or the rain forests in the Amazon survive on their own? With this statement, forestry has finally lost it. In view of the challenges we now face with climate change, more humility would be a fine thing.

YOU CAN SEE FOR YOURSELF just how strong the power of the returning forest is in your own yard or city. Take a look in the flower beds and you'll see tree seedlings popping up everywhere. If you didn't weed your garden, it would become a small woodland within a decade. Birches that sprout in gutters and on tops of walls despite extremely dry summer conditions are also showing their will to survive.

My *aha* moment happened when I was waiting for a group at an event organized by the forest academy. We were to meet at the barbecue shelter in Wershofen, which is located by the community's playing fields. Next to the parking lot was a neglected tennis court clearly no longer in use. It appeared no one had looked after it during the three dry years of 2018, 2019, and 2020, and a multitude of tiny trees had shamelessly taken advantage of the situation. They had settled in by the hundreds, sinking their roots into the dry, compacted sandy surface despite the burning heat of the sun, and they had survived the three record-breaking dry summers just fine. If a new forest was arising even under these extremely unfavorable conditions, I wasn't going to be too worried about the future.

Certainly, we must severely cut back our use of resources and we must finally put an end to blasting enormous amounts of greenhouse gas into the atmosphere. Along with this, we need to give nature more room to put an end to species' extinction. But the question of whether nature and forests can recover on their own is being answered definitively by the stalwart tree children on the tennis court.

These tree children have priceless advantages. They are well adapted to the regional climate and exhibit wide genetic diversity. Seedlings from tree nurseries come from a few recognized seed banks, which means they are the foundation for little forests where the trees grow the way the forest industry wants them to: as thin, straight trees with few sizable branches so their trunks are well suited to being processed into boards and beams. What's most important are visual and technical characteristics. How social are these trees when they are planted out together? How well do they learn? Neither characteristic plays any role in their selection. It reminds me a bit of human intelligence tests that test logical thinking and ignore people skills. Wild trees may not always grow in ways that are optimal for the forest industry, but they are well equipped to survive. This makes them a better choice for us, because the more pressing question in the future will not be how much timber our forests produce but whether we have any forests at all.

ONE QUESTION COMES UP on many of my guided tours of the forest: Can trees left to their wild ways ever successfully create a primeval forest? Or is that impossible? People seeking economic gain have, after all, irretrievably compacted many soils with their harvesters, and this alone has created conditions that make it difficult for forest trees to put down roots. Moreover, many species have died out (mostly tiny ones like bacteria) and cannot be replaced. And even without these constraints, there are no longer any ancient trees, no thick dead trunks. In short: are we chasing a pipe dream?

I don't think so. I think what we need to do is to change our perspective. A primeval forest, even under the most favorable conditions, needs at least one generation of trees that can grow without any active interference from people— that is to say, without any chain saws. Depending on the species, this could be many centuries. That is bad news for impatient beings, which is what we humans are. Add to that the uncertainty about whether such forests could even really grow again—that seems to be asking a lot.

But do we need primeval forests right away? What if we settled for wilderness? The dictionary defines *wilderness* as an impassable, undeveloped, unpopulated place. If we add *unmanipulated*, what we have is… nature! Nature is the opposite of a cultivated landscape; in other words, it is all that we have often gone to a great deal of trouble to change over hundreds of years. As soon as we pull back, the same scenario plays out everywhere, the one I described happening on the tennis court in Wershofen. Forests reclaim their original territory. The longer we leave these areas alone, the wilder they will become.

I greatly prefer the term *wilderness* to the word *nature* because wilderness elicits a much more emotional response—it carries within it the ideas of freedom and adventure. And it is more accurate than official jargon. According to the German Federal Agency for Nature Conservation (BfN), there are 8,833 nature preserves in Germany, covering 6.3 percent of the land. Another category, the Natura 2000 areas (a European network of protected areas), covers even more land in Germany: 15 percent of the country is supposedly set aside to serve nature in these areas.[2] As we've already seen in the example of the ancient beech forests of the Holy Halls, this is not what happens. A similar story can be told about many protected areas, right up to the national parks. The term *nature* is so watered down and misused in this context that the areas only offer on paper what we should expect them to be: areas where people do nothing.

In contrast, we all agree that wilderness really should be left alone. Therefore, it's a good indicator of how much land we really want to set aside for our wild fellow beings. In Germany, this amounts to 0.6 percent of the total land for the year 2020. Only 0.6 percent of the land offers what the other categories promise, namely true protection. The political goal was 2 percent by the year 2020,[3] which also shows that up until now our interests in a number of other categories have been prioritized with little effort made to scale them back.

For example, clear-cuts even larger than those allowed in commercially managed forests are allowed in national parks. The timber is then sold to nearby sawmills, which means the protected area is bleeding valuable biomass.

So pay attention to the term *wilderness*—everything else is mostly a sham.

Wohlleben's Forest Academy's forest conservation project also aims for a change in perspective. Initially, the team stepped up to protect old, partially intact beech forests by leasing them from their owners. The goal was to regenerate primeval forests as quickly as possible. The owners, mostly municipalities, received financial compensation for no longer logging the leased forests. The generous lease contracts replaced contracts with timber companies and not a single tree had to be cut down. The sum per hectare was more than the market price of the trees had the forest been clear-cut. In both scenarios—the lease and the clear-cut—after the initial payment, there would be no more money for the next ten years. In the case of the lease, however, the forest remained intact, the forest owners got paid immediately and could invest their money right away, and none of this was tied to the timber market.

We have since extended the program to all forests. After all, wilderness can also happen in a plantation of dead spruce trees if we allow nature to create a true forest there by itself—as long as the dead spruce trees remain in the forest. There, the dead wood cools the young trees and the bleached trunks still throw some shade. Moreover, both spruce plantations allowed to regrow freely and young deciduous forests provide climate buffer zones around areas where ancient beeches grow.

AND NOW LET'S CIRCLE BACK to the question about the return of the microorganisms that contribute so much to the functioning of the forest ecosystem. For ground dwellers such as beetle mites and springtails, the question is as good as answered, thanks, in fact, to spruce and pine plantations. These trees are not native to most of Europe and therefore we have no naturally occurring species that specialize in living with them. Studies in my forest have found these tiny life-forms even in the areas that were once plantations. Even though the species composition in these former plantations is very different from those in the old, protected beech forests, the little ground dwellers seem content to eat acidic needles. But how did these little guys get to the parts of the forest they now call home?

The most probable answer is animals. Wild boar wallow in mud to free themselves of parasites. While doing so, they acquire some hangers-on, which they offload at another wallow somewhere else. Many beetle mites and springtails don't survive this mode of travel, but luckily there's a more comfortable form of transportation available: birds. The feathered flock love to take dust baths, where they, like the wild boar in their wallows, rid themselves of unwanted guests. To take their dust baths, birds lie on the ground and fluff themselves up. Then they use their wings to distribute dust and humus in between their feathers. They might keep doing this for several minutes. Finally, the birds give themselves a big shake and they fly off into another forest. They, like the wild boar, take a few passengers along with them, which will be offloaded at the next dust bath somewhere else.

Even smaller travelers are bacteria and fungi. Trees are not complete without them. Just think of the holobionts, the ecosystems, trees create (as we do) in combination with thousands of microorganisms. Apart from animals, the tiny travelers have another, even more efficient means of moving from place to place: the wind. Wind picks up the minuscule spores of fungi from the ground and carries them everywhere.

A team led by environmental scientist Bala Chaudhary collected fungal spores on the roof of a five-story building at the DePaul University Lincoln Park Campus in Chicago. After twelve months, she identified more than 47,000 spores from species that cooperate with plant roots underground, which was interesting because underground species have more difficulty spreading their spores than fungi that grow aboveground. Most of the spores she found came from fungi that grow in agricultural fields, where plows churn up both dust and spores.[4]

In forests, of course, there is no plowing. Quite the opposite. Trees use their roots to hold the soil in place to make sure the wind doesn't blow it away. But the fungi are prepared for this and form fruiting bodies that release a multitude of spores to drift on the wind. You can see these spores for yourself. Take a mushroom cap and leave it overnight on a sheet of white paper. When you lift off the cap the next morning, you'll find an impression of the underside of the mushroom created by spores that drifted down overnight.

You're breathing in fungal spores all the time, even now while you're reading this book. Between a thousand and ten thousand spores are drifting around in each 35 cubic feet (each cubic meter) of air—with every breath you take, about ten spores make their way into your lungs.[5]

If fungi from species that grow in primeval forests are to travel to new places, they need one thing above all else: primeval forests. That's why it's so important to preserve the last primeval forests. Where they no longer exist, we must protect the closest thing we have: remnant forests like the Holy Halls in Germany, for example. From these refugia, fungi, bacteria, and all the tiny creatures of the soil can travel to new young forests via airmail. There, they can help the trees rebuild their unique ancient ecosystems.

THE RETURN OF FORESTS can be very exciting and is something we can watch as it happens: nature is change. The farther we pull the pendulum from the ideal, the more momentum it has as it swings back when we let go—in other words, when we let nature do what it wants. And where there's a lot of movement, changes are especially obvious: an agricultural field that is once again covered with young trees within a few years, or a young forest where the poplars and birches grow 40 inches (1 meter) taller every couple of years. These are all things you can follow when you're out walking. Right now, you mostly see the collapse of plantations that have little in common with nature. If we do nothing, what were once green deserts will transform into green wildernesses.

In Germany right now, plantations are the places where the greatest changes are happening from one year to the next. First, the needles of spruce and pines rain down, and everything is transformed into a brown wasteland. Within a year, the ground is covered with grasses, herbs, and thousands of tiny deciduous seedlings. After another year has passed, many little deciduous trees tower over the other plants and begin to shade the ground. Five to ten years later, a young forest covers the whole area. Grasses, herbs, and bushes gradually disappear as it becomes too shady for them. Under the birches and poplars, here and there, beeches, oaks, and maples invite themselves in. They catch up with the early arrivals, overtake them, and within a few decades are running the show.

If you want to follow what's happening near you, I recommend taking photographs from the same position at regular intervals. You could take them from a fork in the trail or a particular viewpoint that you'll be able to recognize years later. Trees are slow, but in your series of photographs you will quickly notice how nature changes.

What's the point of all this? To motivate us. If we personally experience how things are changing for the better, we gather fresh courage to face the future. I'm not stirring up optimism for optimism's sake. We have every reason to hope that forests can cope with the challenges they face because of us. What's important is that we finally accept that the trees themselves know best how to rebuild their ancient ecosystems.

SCIENTISTS RECENTLY NAMED a new geological era: the Anthropocene. We need to end this era. Not in the sense that either we or the societies we have created should disappear. Instead, we need to rejoin the cycle of nature and allow our fellow creatures the space they need so that they, too, can calmly look to the future. The widespread return of the forests that once covered most continents would be a hopeful sign for the future. I've sketched out how this might be possible based on a reduction of meat consumption, for example. I'm hoping that soon we might be able to name a new era: the era of the trees.

I would like to end this book by referring back to the title of this chapter and expanding with words from the film *The Hidden Life of Trees*. Finishing the phrase makes it clear that we must shift our focus: "The forest will return. It would just be nice if we were still around."

Afterword

ACCEPTING IGNORANCE
AND TREADING
CAREFULLY IN THE FOREST

HUMAN-CAUSED CLIMATE CHANGE is messing things up—to put it mildly. It is threatening the survival of the world as we know it. A few decades ago, when scientists began thinking about what human-caused warming of the planet fueled by greenhouse gases might mean for nature, the risks were still abstract. We had no idea what was coming. Over the past few years, however, what was once abstract is now real. Forests are in crisis in many areas of the world: forested landscapes are drying out, wildfires are increasing, hundred-year-old trees are suddenly dying because they cannot withstand droughts that last for years, hot dry air is damaging sensitive plant tissues, and animals are suffering from high temperatures and lack of water and food.

Climate change is causing stress for people and nature. It is also upsetting the disciplines of natural and forest sciences as never before. People expect responses from scientists to questions for which there are no good answers. What will happen? What will forests look like in the future? What changes can we make now to be better prepared for future challenges? Suddenly, it's no longer simply a matter of undertaking new research and communicating proven facts, but of figuring out how best to deal with the fact that we are living in times of great uncertainty. Foresters plan for the long term. And forestry has always been about betting on the future. And that was fine, as long as we could work on the assumption that the future would not be that much different from the present.

Scientists have learned to measure and describe things as accurately as possible. They sort elements of nature according to their form, their provenance, and their function. Researchers find natural laws and rules, which they use to explain why certain phenomena exist. For example, generations of forest scientists have studied how trees grow and how much wood they produce in their lifetimes. Using yield tables, foresters can calculate when and how much they can harvest. Evaluating local conditions and their suitability for various tree species is important to foresters as they plan the future of their forests. In digital times, we now have computers that can run the calculations and supposedly give more accurate answers to these questions. The models, however, are only as accurate as the data fed into them. If an important factor in the model is overlooked or perhaps even unknown, the results quickly become irrelevant. You could have measured and documented how particular trees grew in the past—but if the climate in the future is different, then experiences and formulas from the past are no longer useful.

Climate change draws a big fat line through our calculations. Suddenly, there are completely new formulas for plant growth, to take just one example. It's gradually dawning on us that in a few decades, local growing conditions could be completely different—much hotter and drier, for example. So different that many of the plants and animals we are familiar with will perhaps no longer be able to survive in their ancestral homes. Perhaps! But when will it come to this? Do we have to act now?

In the past, foresters did not have to worry whether the trees they cared for or planted would still be there to be harvested by their successors in 100 years' or 120 years' time. Today, the situation is completely different. We know how important it is to look into the future, but the view is much foggier than it used to be. We have been researching nature for centuries now, redoubling our efforts and using increasingly precise instruments and methods—and yet we have to acknowledge that we cannot answer even the simplest questions. What is going to happen? We do not know. Our ignorance is not simply about gaps in our knowledge that researchers could fill with a little more effort. This ignorance is insoluble; it cannot be fixed. All we can do is learn to live with it.

STARTING IN 2018, when extreme summer storms began to visibly stress forests, and as increasing numbers of trees died and in some areas swathes of land turned from green to brown, television, radio, and newspaper reporters wanted to know what could be done to save the forests. "How sick are the forests?" "Are we experiencing a new forest die-off?" "What trees should be planted now?" And the same question over and over again: "What will forests look like in the future?" Politicians asked the same questions. It's an uncomfortable situation for scientists, because members of the media and decision-makers expect short, clear answers. They do not want to hear "maybe or maybe not" and certainly not "I don't know."

Those who do not give simple answers will likely not be asked again. There is a great temptation, therefore, to say things will happen a certain way and to give concrete recommendations. An approach taken by some scientists, for example, is to recommend specific tree species that show promise for the future—these are usually species from other continents such as Douglas firs, red oaks, or Japanese larches. As a result, these species are mass-planted in forests. It is not at all clear, however, whether these species will deal well with the climate in the future and whatever the conditions turn out to be. It is unclear whether the supposed supertrees will successfully integrate into ecosystems or whether they will be ravaged by disease, and it is possible that planting so many new trees will further weaken natural forests.

Currently, outbreaks of diseases such as ash dieback and sooty bark disease in maples, and infestations of insects such as various moths and bark beetles, are putting pressure on many species of trees. The trees are particularly susceptible if they have previously been weakened by conditions such as drought or heat. In the past, foresters and scientists were constantly surprised by where, when, and which species suffered damage. Not a single one of these species-specific crises was reliably predicted. It is not even possible to make these predictions because so many individual factors influence each other. Essentially, all we could and can be certain of is that the threats to forests and everything that lives in them are increasing hugely because of climate change. We cannot be certain of anything else. That in itself is a big problem, and to act as though this uncertainty does not exist is dangerous.

The all-too-common practice of presenting what we cannot be sure of as fact is also risky. Some scientists, for example, produce colorful maps based on computer models that show which tree species might grow where in the future and which might encounter difficulties. The time periods are often in the second half of this century, for example 2041–2070. The time spans alone suggest certainty where none exists. More importantly, these calculations are based on specific climate conditions that arise given a particular set of circumstances. Yet the climate is teaching us that it can constantly surprise us with something new at short notice. Unfortunately, today's lessons regularly remind us that we have underestimated the extent and uncertainty of climate change.

No one could have predicted that the month of April in the last decade in Germany would suddenly change so completely and become unusually dry and warm. We had no idea that the jet stream would bring us record-breaking weather in summer. We did not really even know that the jet stream existed or that it influenced our weather. No model warned us about a multiyear drought in vast parts of Germany. Even recently, very few forest scientists anticipated the kind of forest crisis we are now experiencing. The problem was—in the truest sense of the word—incalculable.

WHAT LIES AHEAD for us and for forests? And is there anything we can do? The situation can be compared to driving a car in the mountains along what we have to assume is an extremely dangerous stretch of road. We have never driven this road before. We expect there will be sharp curves, steep drop-offs, and traffic that will suddenly appear from the opposite direction in very narrow spots without guardrails. Landslides and rockfalls could be a threat—especially in rainy weather, when the road will also be slippery and fog will restrict visibility.

Let's imagine three different types of driver. A risk-taker would say that they have never had an accident when driving—and never will have one—and race off. In this case it is clear that past experience will not necessarily be helpful on the stretch on which the driver is about to embark. A driver who puts their faith in technology would, among other things, check the latest weather reports before setting off, try to find out something about traffic conditions, and equip the vehicle with airbags, steering and brake assistance systems, and warning lights. The careful and risk-averse driver, in contrast, would certainly pay attention to technological safety measures and buckle up, but the most important things this driver would do would be to drive slowly and be prepared for unexpected traffic coming from the other direction at every curve: they would honk their horn at appropriate intervals and regularly ready themselves to come to a screeching halt if that's what they need to do.

Let's now apply this to the forest. First, the mantra "Everything has always worked out well, so it's business as usual" is no longer an option. Second, more knowledge and more technology will not protect us from uncontrollably rapid changes around the next corner. The only thing we are left with is to proceed with caution and be prepared. The most immediate task is for us to be clear that the future hides dangers we cannot even imagine and to accept that we will be surprised. We should get as much information as we can about the dangers that loom before us, but there's no point trying to predict the unpredictable. What we need to do is slow down.

In the forest, this means not demanding so much from it and manipulating and exploiting it less. Instead, we should be fostering the strengths that make it resilient. When it's more than likely that the climate is going to get hot and dry with extremes, then it's less important to know exactly when which conditions might prevail and whether it's going to get two or three degrees warmer than it was 150 years ago. It's far more important to support forests so they can keep themselves as cool and moist as possible.

WE KNOW FOREST ECOSYSTEMS are, at their heart, complex superorganisms that are healthier when the networks that connect all the elements are intact. Therefore, it's less important to spend a lot of time studying and describing all these parts and how they are connected and more important to ensure that they are not destroyed.

We don't know whether forest ecosystems are strong enough to adapt to the challenges they face. But why do we think we have a better solution for the problem simply because we are able to figure out in advance that there will be a problem? Nature is not a simple clockwork mechanism that ticks at the same rate in perpetuity. Rather, forests are complex systems that process information. They are places where information about solutions to problems are stored in the genetic material of organisms and in their interplay with each other. As the forest evolves, this knowledge is constantly being tested and added to. This means you can, literally, talk of ecosystem intelligence. This intelligence has no need of consciousness or even the ability to imagine the future, but it is exactly what nature needs to react to unpredictable events.

When, for example, a forest burns, there are pioneer species whose seeds arrive quite quickly to restart the ecosystem. They are already adapted to dealing with the difficult conditions that dominate freshly burned areas. They can germinate without humus and tolerate extreme chemical and physical conditions. These trees, quaking aspen, for example, often find that the important mycorrhizal partners they rely on to get a better start in life are already waiting for them in the burn area.

None of this has to be reinvented after a fire. The ecosystem can summon the solution—completely unconsciously—from its "memory." That is how forest ecosystems heal "wounds" inflicted by unpredictable disturbances such as fires and storms. The vegetation fills in and in no time at all, the ground is covered and protected. The activity of pioneer species ensures that new soil forms and that this soil is shaded and cooled. Desperately needed water is retained and increasing numbers of species arrive to rebuild the ecosystem. It does not seem too much of a stretch to call processes such as these the "self-healing" powers of nature.

Similar processes and capabilities have been described for many different ecosystems. They also come into effect when a change in the environment or a disease causes important individual species to die out. You can see that today in the forest crisis. In places where, for example, beeches in extremely demanding locations died after prolonged drought, the forest itself did not die. Other, more drought-tolerant species such as hornbeams and lindens got the chance to become part of the regeneration of the forest.

TO JUSTIFY THE IDEA that foresters faced with climate change must now intervene even more actively than before—and should on no account trust natural process—a well-known professor of silviculture from Freiburg said in an interview with the *Süddeutsche Zeitung* that the idea that nature could heal itself was "completely lacking in evidence." Lack of evidence is one of the most serious accusations a scientist can level against their colleagues. "Completely lacking in evidence" means there is nothing to back up the scientist's statements or recommendations—it means, therefore, that the ideas the scientist is putting forward are unscientific.

In this case, the accusation appeared to uncover a twofold crisis in forest sciences. On the one hand, it showed that current knowledge about what is happening in nature is being ignored or actively denied to justify what forestry is doing. On the other, it pointed to a serious scientific misunderstanding. In these desperate times, scientists are no longer the ultimate authorities that provide evidence and make recommendations. After all, who can prove that the ecosystems of the future can deal with all the challenges they now face? No one, of course—and if climate change turns out to be as intense as credible scenarios suggest, all the evidence indicates ecosystems will fail. Indeed, everything turns on this issue of credibility. How credible and probable is it that forest scientists who just a few years ago were unable to predict the severe forest crisis we now find ourselves in, suddenly have better ideas about developing forests for the future than nature itself, which has been in training for millions of years in how to deal with the unknown and surprising?

Climate change teaches us to be humble in this respect as well. We desperately need to learn how we can manage our ignorance better. We should not be so sure of ourselves in everything—and we should not so casually dismiss what nature knows. Instead of believing that smart engineers with their technological solutions will save us, we should look to the tried-and-true principles of being cautious and being prepared. If we accept and respect our ignorance, it can set us on the right path.

PROFESSOR PIERRE IBISCH

Acknowledgments

I HAVE WRITTEN many books and dedicated many of them to my family. I have also thanked my publishers and the staff there who have worked with me. In this book, I would like to bring Lars Schultze-Kossack out of the shadows of his office. He works modestly and yet tremendously effectively in representing my books. Lars, his wife Nadja, and the whole team at the agency handle contracts, field requests, defend my work against copyright infringements, and even helped get a documentary up and running. Business negotiations are not my thing—I much prefer to hand everything over to someone else. It's great, then, that Lars can set the necessary boundaries and yet, most importantly, keep the door open to opportunity. Without him, I would never have ended up with Ludwig as my publisher, a place where I feel very much at home.

I would also like to thank my coworkers at the forest academy. They are the point of contact for all the enthusiastic readers who still have questions or simply want to drop by to find out where all the trees are that have inspired me so much. As the team does all the work, I can concentrate on being an educator. In this role, I have the pleasure of taking people out into the forests of the Eifel and doing what I like to do best: explaining how trees live their lives.

Notes

1 | When Trees Make Mistakes

1. Katrin Blawat, "Warum gerade viele Kastanienbäume blühen," *Süddeutsche Zeitung*, October 3, 2020, www.sueddeutsche.de/wissen/kastanien-schaedlinge-blueteumwelt-1.5052988.

2. For example, "Kurios im Herbst: Blühende Bäume schmücken die Natur," inFranken.de, September 5, 2018, www.infranken.de/ratgeber/garten/gartenjahreszeiten/kurios-im-herbst-bluehende-baeume-schmuecken-die-naturin-franken-art-3666516.

3. Thomas Samboll, "Deshalb hilft es Pflanzen, wenn wir mit ihnen reden," swr, April 8, 2020, www.swr.de/wissen/haben-pflanzen-gefuehle-100.html.

4. "Haben Pflanzen ein 'Gehirn'?" Bloomling, www.bloomling.de/info/ratgeber/haben-pflanzen-ein-gehirn.

5. Frank Hagedorn, Jobin Joseph, Martina Peter, et al., "Recovery of Trees From Drought Depends on Belowground Sink Control," *Nature Plants* 2, no. 16111 (July 18, 2016), doi.org/10.1038/nplants.2016.111.

6. Emily F. Solly, Ivano Brunner, Heljä-Sisko Helmisaari, et al., "Unravelling the Age of Fine Roots of Temperate and Boreal Forests," *Nature Communications* 9, no. 3006 (August 1, 2018), doi.org/10.1038/s41467-018-05460-6.

2 | A Thousand Years of Learning

1. Beth Gibson, Daniel J. Wilson, Edward Feil, and Adam Eyre-Walker, "The Distribution of Bacterial Doubling Times in the Wild," *Proceedings of the Royal Society B* 285, no. 1880 (2018): 20180789, doi.org/10.1098/rspb.2018.0789.

2. Jörn Auf dem Kampe, "Man kann die Erbse trainieren, fast wie einen Hund," geo, September 2019, m.geo.de/natur/naturwunder-erde/

21836-rtkl-kluge-pflanzen-man-kann-die-erbse-trainieren-fast-wie-einen-hund.

3. Tourismusverband Mecklenburgische Seenplatte, "Nationales Natur-monument Ivenacker Eichen," www.mecklenburgische-seenplatte.de/reiseziele/nationales-naturmonument-ivenacker-eichen.

4. Katharina Weltecke, Jörn Benk, Jonas Heck, et al., "Rätsel um die älteste Ivenacker Eiche," AFZ, der Wald, no. 24 (2020): 12–17, standort-baum.de/media/afz_24_20_weltecke_raetsel_um_ivenacker_eichen.pdf.

5. Andreas Roloff, "Vitalität der Ivenacker Eichen und baumbiologische Überraschungen," AFZ, der Wald, no. 24 (2020): 18–21.

3 | Seeds of Wisdom

1. Carl Zimmer, "The Famine Ended 70 Years Ago, but Dutch Genes Still Bear Scars," New York Times, January 31, 2018, www.nytimes.com/2018/01/31/science/dutch-famine-genes.html.

2. Technical University of Munich, "An Epigenetic Ageing Clock in Trees," Research News, November 18, 2020, www.tum.de/en/about-tum/news/press-releases/details/36315/.

3. Arun K. Bose, Barbara Moser, Andreas Rigling, et al., "Memory of Environmental Conditions Across Generations Affects the Acclimation Potential of Scots Pine," Plant, Cell & Environment 43, no. 5 (May 2020): 1288–1299, doi.org/10.1111/pce.13729.

4. Erwin Hussendörfer, "Baumartenwahl im Klimawandel: Warum (nicht) in die Ferne schweifen?!" in Der Holzweg (Munich: Oekom Verlag, 2021), 222.

4 | Filling Up in Winter

1. Scott T. Allen, James W. Kirchner, Sabine Braun, et al., "Seasonal Origins of Soil Water Used by Trees," Hydrology and Earth Systems Sciences 23, no. 2 (2019): 1199–1210, doi.org/10.5194/hess-23-1199-2019.

2. Kompetenz- und Informationszentrum Wald und Holz, "Wald und Boden—Wasserfilter und Wasserspeicher," www.kiwuh.de/service/wissenswertes/wissenswertes/wald-boden-wasserfilter-wasserspeicher.

3. Umweltbundesamt, "Veränderung der jahreszeitlichen Entwicklungs-phasen bei Pflanzen," May 2, 2021, www.umweltbundesamt.de/daten/klima/veraenderung-der-jahreszeitlichen#pflanzen-als-indikatoren-furklimaveranderungen.

4. Lothar Zimmermann, Stephan Raspe, Christoph Schulz, and Winfried Grimmeisen, "Wasserverbrauch von Wäldern," in Wald und Wasser, LWF

aktuell 66 (2008): 16, www.lwf.bayern.de/mam/cms04/boden-klima/dateien/a66-wasserverbrauch-von-waeldern.pdf.

5. R. C. Ward and M. Robinson, *Principles of Hydrology*, 3rd ed. (Maidenhead: McGraw-Hill, 1989).

6. Johann Siedl, "Laubfall im Herbst," press release, Bayerischen Landesanstalt für Wald und Forstwirtschaft, www.lwf.bayern.de/service/presse/089262/index.php.

7. MartinFlade and Susanne Winter, "Wirkungen von Baumartenwahl und Bestockungstyp auf den Landschaftswasserhaushalt," in *Der Holzweg* (Munich: Oekom Verlag, 2021), 240.

5 | Red Flags for Aphids

1. Taylor S. Feild, David W. Lee, and N. Michele Holbrook, "Why Leaves Turn Red in Autumn. The Role of Anthocyanins in Senescing Leaves of Red-Osier Dogwood," *Plant Physiology* 127, no. 2 (October 2001): 566–574, doi.org/10.1104/pp.010063.

2. W. D. Hamilton and S. P. Brown, "Autumn Tree Colours as a Handicap Signal," *Proceedings of the Royal Society* B 268, no. 1475 (July 2001): 1489–1493, doi.org/10.1098/rspb.2001.1672.

3. Thomas F. Döring, "How Aphids Find Their Host Plants, and How They Don't," *Annals of Applied Biology* 165, no. 1 (June 2014): 3-26, doi.org/10.1111/aab.12142.

4. Marco Archetti, "Evidence From the Domestication of Apple for the Maintenance of Autumn Colours by Coevolution," *Proceedings of the Royal Society* B 276, no. 1667 (July 2009): 2575–2580, doi.org/10.1098/rspb.2009.0355.

5. Deborah Zani, Thomas W. Crowther, Lidong Mo, Susanne S. Renner, and Constantin M. Zohner, "Increased Growing-Season Productivity Drives Earlier Autumn Leaf Senescence in Temperate Trees," *Science* 370, no. 6520 (November 2020): 1066–1071, doi.org/10.1126/science.abd8911.

6 | Early Risers and Late Sleepers

1. Volker Mrasek, "Winter in Deutschland werden immer warmer," Deutschlandfunk, December 21, 2020, www.deutschlandfunk.de/klimawandel-winter-in-deutschland-werden-immer-waermer-100.html.

2. "Bäume spüren den Frühling," *Schweriner Volkszeitung*, March 25, 2019.

3. Michèle Kaennel Dobbertin, "War der letzte Winter zu warm für unsere Waldbäume?" press release, Eidgenössische Forschungsanstalt

für Wald, Schnee und Landschaft WSL, March 19, 2020, www.wsl.ch/
de/202%3/war-der-letzte-winter-zu-warm-fuer-unsere-waldbaeume
.html.

7 | Forest Air-Conditioning

1. "Gericht stoppt vorläufig Rodung im Hambacher Forst," *Spiegel*,
   October 5, 2018, www.spiegel.de/wirtschaft/soziales/hambacher-
   forst-gericht-verfuegt-einstweiligen-rodungs-stopp-a-1231705.html.
2. Pierre L. Ibisch, Steffen Kriewald, and Jeanette S. Blumröder, "Hambacher Forst in der Krise: Studie zur Beurteilung mikro- und mesok
   limatischen Situation sowie Randeffekten," www.greenpeace.de/
   publikationen/hambacher_forst.pdf.
3. Greenpeace, "Hitze Sichtbar Gemacht," August 13, 2020, www.green
   peace.de/klimaschutz/energiewende/kohleausstieg/hitze-sichtbar.
4. Ministeriums für Klimaschutz, Umwelt, Energie, und Mobilität,
   "Landesforsten RLP: Einschlagstopp für alte Buchen im Staatswald,"
   September 3, 2020, mkuem.rlp.de/de/pressemeldungen/detail/
   news/News/detail/landesforsten-rlp-einschlagstopp-fuer-alte-
   buchen-im-staatswald/.

8 | When Rain Falls in China

1. Lothar Zimmermann, Stephan Raspe, Christoph Schulz, and Winfried
   Grimmeisen, "Wasserverbrauch von Wäldern," *LWF aktuell* 66 (2008):
   16–20.
2. Anastassia M. Makarieva and Victor G. Gorshkov, "Biotic Pump of
   Atmospheric Moisture as Driver of the Hydrological Cycle on Land,"
   Hydrology and Earth Systems Sciences 11, no. 2 (2007): 1013–1033,
   doi.org/10.5194/hess-11-1013-2007.
3. Ulrich Weihs, "Unterscheiden sich Laubbäume in ihrer Anpassung
   an Trockenheit? Wie viel Wasser brauchen Laubbäume?" Max-
   Planck-Institut für Dynamik und Selbstorganisation, www.ds.mpg
   .de/139253/05.
4. Douglas Sheil, "Forests, Atmospheric Water and an Uncertain Future:
   The New Biology of the Global Water Cycle," *Forest Ecosystems* 5, no. 19
   (2018), doi.org/10.1186/s40663-018-0138-y.
5. Rudi J. van der Ent, Hubert H. G. Savenije, Bettina Schaefli, and Susan
   C. Steele-Dunne, "Origin and Fate of Atmospheric Moisture Over
   Continents," *Water Resources Research* 46, no. 9 (September 2010),
   doi.org/10.1029/2010WR009127.

6. Bernd Dörries, "Kampf ums Wasser," *Süddeutsche Zeitung*, June 29, 2020, www.sueddeutsche.de/politik/aegypten-aethiopien-nil-damm-1.4950300.
7. Alexander von Humboldt, *Fragments of a Geology and Climatology of Asia* (Berlin: J. A. List, 1832), 239.

9 | Take Care and Stand Back

1. "Arabidopsis thaliana," Spektrum.de, www.spektrum.de/lexikon/biologie-kompakt/arabidopsis-thaliana/815.
2. María A. Crepy and Jorge J. Casal, "Photoreceptor-Mediated Kin Recognition in Plants," *New Phytologist* 205 (2015): 329–338, doi.org/10.1111/nph.13040.
3. Kathleen J. Wu, "Some Trees May 'Social Distance' to Avoid Disease," *National Geographic*, July 6, 2020, www.nationalgeographic.com/science/article/tree-crown-shyness-forest-canopy.
4. Roza D. Bilas, Amanda Bretman, and Tom Bennett, "Friends, Neighbours and Enemies: An Overview of the Communal and Social Biology of Plants," *Plant, Cell & Environment* 44, no. 4 (April 2021): 997–1013, doi.org/10.1111/pce.13965.

10 | Underrated All-Rounders

1. Kelly S. Ramirez, Jonathan W. Leff, Albert Barberán, et al., "Biogeographic Patterns in Below-Ground Diversity in New York City's Central Park Are Similar to Those Observed Globally," *Proceedings of the Royal Society B* 281, no. 1795 (November 22, 2014), doi.org/10.1098/rspb.2014.1988.
2. R. J. Rodriguez, J. F. White Jr., A. E. Arnold, and R. S. Redman, "Fungal Endophytes: Diversity and Functional Roles," *New Phytologist* 182, no. 2 (April 2009): 326, doi.org/10.1111/j.1469-8137.2009.02773.x.
3. Martin Hubert, "Der Mensch als Metaorganismus," Deutschlandfunk, December 30, 2018, www.deutschlandfunk.de/meine-bakterien-und-ich-der-mensch-als-metaorganismus-100.html.
4. "Did Nerve Cells Evolve to Talk to Microbes?" press release, Kiel University, July 10, 2020, www.uni-kiel.de/en/university/details/news/168-klimovich-pnas.
5. EEA Report no. 10/2020, p. 16, box 2.1, www.eea.europa.eu/publications/state-of-nature-in-the-eu-2020.
6. Noah Fierer, Micah Hamady, Christian L. Lauber, and Rob Knight, "The Influence of Sex, Handedness, and Washing on the Diversity of Hand

Surface Bacteria," PNAS 105, no. 46 (November 18, 2008): 17994–17999, doi.org/10.1073/pnas.0807920105.

7. Joachim Schüring, "Wie viele Zellen hat der Mensch?" Spektrum.de, www.spektrum.de/frage/wie-viele-zellen-hat-der-mensch/620672.

8. Pierre L. Ibisch and Jeanette S. Blumröder, "Waldkrise als Wissenskrise als Risiko," Universitas 888 (2020): 20–42.

9. Heribert Cypiomka, "Von der Einfalt der Wissenschaft und der Vielfalt der Mikroben," www.pmbio.icbm.de/download/einfalt.pdf.

10. Sophie Nguyen, Kristi Baker, Benjamin Padman, et al., "Bacteriophage Transcytosis Provides a Mechanism to Cross Epithelial Cell Layers," mBio 8, no. 6 (November 21, 2017), doi.org/10.1128/mBio.01874-17.

11. Gijsbert D. A. Werner, William K. Cornwell, Janet I. Sprent, et al., "A Single Evolutionary Innovation Drives the Deep Evolution of Symbiotic N_2 Fixation in Angiosperms," Nature Communications 5, no. 4087 (June 10, 2014), doi.org/10.1038/ncomms5087.

12. Jos M. Raaijmakers and Mark Mazzola, "Soil Immune Responses," Science 352, no. 6292 (June 17, 2016): 1392–1393, doi.org/10.1126/science .aaf3252.

13. "Jährliche Änderung des Waldbestandes," Bundeszentrale für politische Bildung, September 1, 2017, www.bpb.de/nachschlagen/ zahlen-und-fakten/globalisierung/52727/waldbestaende.

12 | Butchery in the Beech Forest

1. "Erste Baumsprengung in Thüringen stellt Experten vor Probleme," Thüringer Allgemeine, September 8, 2019, www.thueringer-allgemeine.de/ leben/natur-umwelt/die-ersten-baeume-werden-dieses-wochenende-in-thueringen-gesprengt-id227017755.html.

2. Bundesgerichtshof VI-ZR 311/11, "Eine Haftung des Waldbesitzers wegen Verletzung der Verkehrssicherungspflicht besteht grundsätzlich nicht für waldtypische Gefahren," October 2, 2012, www.dstgb.de/ themen/kommunalwald/aktuelles/urteil-des-bgh-zur-verkehrssi cherungspflicht-im-wald/anlage-12-bgh-urt-verkehrssicherungspflicht-v-2-10-2012-vi-zr-311-11.pdf?cid=90i.

3. "Deutliches Ergebnis: Nadelholz ist nicht ersetzbar," Holz-Zentralblatt 18 (April 30, 2015): 391, waldmaerker.de/pdf/downloads/Nadelholz_ist_ nicht_ersetzbar.pdf.

13 | Germany's Search for the Supertree

1. For example, Reinhard Zweigler, "Wegen des Klimawandels: 'Pakt für den Wald' schließen," *Märkische Allgemeine*, March 24, 2019, www.maz-online.de/Brandenburg/Wegen-des-Klimawandels-Pakt-fuer-den-Wald-schliessen.

2. Karl von Koerber, Jürgen Kretschmer, and Stefani Prinz, "Globale Ernährungsgewohnheiten und -trends," commissioned expert study for the German Advisory Council on Global Change (WBGU) Report, *World in Transition: Future Bioenergy and Sustainable Land Use* (Berlin: WBGU, December 8, 2008), www.wbgu.de/en/publications/publication/welt-im-wandel-zukunftsfaehige-bioenergie-und-nachhaltige-landnutzung#section-downloads.

3. Joachim Rock and Andreas Bolte, "Welche Baumarten sind für den Aufbau klimastabiler Wälder auf welchen Böden geeignet? Eine Handreichung," Thünen-Institut für Waldökosysteme, February 2, 2020, www.waldbodenschutz.de/images/pdf/rock-bolte-thuenen-institut_2020.pdf.

4. Axel Vogel, "Rheinbacher Wald in katastrophalem Zustand," *General Anzeiger*, September 4, 2018, ga.de/region/voreifel-und-vorgebirge/rheinbach/rheinbacher-wald-in-katastrophalem-zustand_aid-43889517.

5. Stefan Huber and Karin Bork, "Blattfraß an Baumhasel durch die Breitfüßige Birkenblattwespe," *AFZ-Der Wald*, October 21, 2020, www.forstpraxis.de/blattfrass-an-baumhasel-durch-die-breitfuessige-birkenblattwespe/.

6. "Können Bäume eine schwere Grippe bekommen?" press release, Humboldt-Universität zu Berlin, August 6, 2020, www.hu-berlin.de/de/pr/nachrichten/august-2020/nr-2086.

14 | Good Intentions, Poor Outcomes

1. Bauhaus, "Weil es richtig wichtig ist, pflanzen wir 1 Million Bäume," richtiggut.bauhaus.info/1-million-baeume/initiative.

2. Bauhaus, "Ihr habt Fragen zur BAUHAUS Klimawald Initiative?" richtiggut.bauhaus.info/1-million-baeume/initiative/faq.

3. Schutzgemeinschaft Deutsche Wald, "Unser Leitbild," www.sdw.de/ueber-die-sdw/unser-leitbild/.

4. Schutzgemeinschaft Deutscher Wald, "Pflanzkodex der Schutzgemeinschaft Deutscher Wald im Rahmen der Kooperation mit BAUHAUS," May 2020, www.sdw.de/fileadmin/Bundesverband/PDF_Dokumente/Pflanzkodex_Bewerbungsbogen.pdf.

5. "Langfristig sind reale Renditen entscheidend," Growney, October 16, 2017, growney.de/blog/langfristig-sind-reale-renditen-entscheidend.

6. Eight hundred cubic meters of wood after 100 years, of which no more than 400 cubic meters will be high-value timber, which—after you have deducted the costs of woodlot management and harvesting—will bring in an average of 30 euros per cubic meter, for a total of 12,000 euros.

15 | The New Bark Beetle?

1. Frank Tottewitz, Grit Greiser, Ina Martin, and Johanna M. Arnold, "Streckenstatistik in Deutschland—ein wichtiges Instrument im Wildtiermanagement," 2016 annual conference, Gesellschaft für Wildtier- und Jagdforschung (GWJF), www.wild-auf-wild.de/sites/default/files/1-WILD_PosterGWJF_2016_Jagdstrecke.pdf.

16 | Wolves as Climate Guardians

1. "Wolf occurrence," Dokumentations und Beratungsstelle des Bundes zum Thema Wolf (Federal Documentation and Consultation Center on Wolves), www.dbb-wolf.de/wolf-occurrence/confirmed-territories/map-of-territories.

2. "Was frisst der Wolf?" NABU (Naturschutzbund Deutschland), www.nabu.de/tiere-und-pflanzen/saeugetiere/wolf/wissen/15572.html.

3. "Was ist ein Territorium und wie groß ist es?" Dokumentations- und Beratungsstelle des Bundes zum Thema Wolf (Federal Documentation and Consultation Centre on Wolves), www.dbb-wolf.de/mehr/faq/was-ist-ein-territorium-und-wie-gross-ist-es.

4. Felix Knauer, Georg Rauer, and Tanja Musil, "Der Wolf kehrt zurück—Bedeutung für die Jagd?" Weidwerk 9 (2016): 18–21.

5. Selwyn Hoeks, Mark A. J. Huijbregts, Michela Busana, et al., "Mechanistic Insights Into the Role of Large Carnivores for Ecosystem Structure and Functioning," Ecography 43, no. 12 (December 2020): 1752–1763, doi.org/10.1111/ecog.05191.

17 | Is Wood as Eco-Friendly as We Think?

1. One of many examples: "10 Gründe mit Holz zu heizen," Landesforsten, Rheinland-Pfalz, www.wald.rlp.de/de/forstamt-trier/angebote/brennholz/10-gruende-mit-holz-zu-heizen/.

2. Hans Pretzsch, "The Course of Tree Growth. Theory and Reality," Forest Ecology and Management 478 (2020): 118508, doi.org/10.1016/j.foreco.2020.118508.

3. Federal Ministry of Food and Agriculture, *The Forests in Germany: Selected Results of the Third National Forest Inventory* (Berlin: Federal Ministry of Food and Agriculture, 2015), 16.

4. Gianluca Piovesan, Franco Biondi, Michele Baliva, et al., "Lessons From the Wild: Slow but Increasing Long-Term Growth Allows for Maximum Longevity in European Beech," *Ecology* 100, no. 9 (September 2019): e02737, doi.org/10.1002/ecy.2737.

5. A. Frühwald, C. M. Polmann, G. Wegener, *Holz—Rohstoff der Zukunft nachhaltig verfügbar und umweltgerecht*, Informationsdienst Holz (Deutsche Gesellschaft für Holzforschung and Holzabsatzfonds: Munich, 2001), silo.tips/download/holz-informationsdienst-holz-rohstoff-der-zukunft-nachhaltig-verfgbar-und-umwelt.

6. Bundesministerium für Ernährung und Landwirtschaft, Fachagenture Nachwachsende Rohstoffe, *Rohstoffmonitoring Holz: Daten und Botenschaften*, www.fnr.de/fileadmin/allgemein/pdf/broschueren/Handout_Rohstoffmonitoring_Holz_Web_neu.pdf.

7. Jana Ballenthien and Ronja Heise, "Aktionstag: Wilde Wälder schützen—nicht verfeuern!" Robinwood, November 24, 2020, www.robinwood.de/blog/aktionstag-wilde-w%C3%A4lder-sch%C3%BCtzen-%E2%80%93-nicht-verfeuern.

8. "Letter From Scientists to the EU Parliament Regarding Forest Biomass," updated January 14, 2018, www.pfpi.net/wp-content/uploads/2018/04/UPDATE-800-signatures_Scientist-Letter-on-EU-Forest-Biomass.pdf.

9. Sebastian Rüter, Frank Werner, Nicklas Forsell, et al., "ClimWood2030, Climate Benefits of Material Substitution by Forest Biomass and Harvested Wood Products: Perspective 2030—Final Report," Johann Heinrich von Thünen-Institut, Report 42 (2016): 106, www.thuenen.de/media/publikationen/thuenen-report/Thuenen_Report_42.pdf.

10. "Klima: Der große Kohlenspeicher," Heinrich Böll Stiftung, January 8, 2015, www.boell.de/de/2015/01/08/klima-der-grosse-kohlenspeicher.

11. J. Reise, C. Urrutia, H. Böttcher, and K. Henneberg, "Literaturstudie zum Thema Wasserhaushalt und Forstwirtschaft," *Öko-Institut* (September 2020): 12, www.oeko.de/publikationen/p-details/literaturstudie-zum-thema-wasserhaushalt-und-forstwirtschaft.

12. Christopher Dean, Jamie B. Kirkpatrick, Richard B. Doyle, et al., "The Overlooked Soil Carbon Under Large, Old Trees," *Geoderma* 376 (October 15, 2020): 114541, doi.org/10.1016/j.geoderma.2020.114541.

13. Rainer Soppa, "Waldbauern fordern 5% aus CO_2-Abgabe als Anerkennung für die Klimaschutzleistung ihrer Wälder," *Forstpraxis*, November 17, 2020, www.forstpraxis.de/waldbauern-fordern-5-aus-co2-abgabe-als-anerkennung-fuer-die-klimaschutzleistung-ihrer-waelder/.

18 | It's Time to Pay Up

1. Christian Meyer, "Für diese Technologie will Elon Musk einen Millionenpreis vergeben," *Frankfurter Allgemeine Zeitung*, January 22, 2021, www.faz.net/aktuell/wirtschaft/co2-bindung-elon-musik-vergibt-preis-fuer-diese-technologie-17159260/tesla-chef-elon-musk-17159342.html.

2. "Carbon Capture and Storage," Umweltbundesamt, January 15, 2021, www.umweltbundesamt.de/themen/wasser/gewaesser/grundwasser/nutzung-belastungen/carbon-capture-storage#grundlegende-informationen.

3. Marcus Fairs, "Norway Begins Work on 'Absolutely Necessary' Project to Bury up to 1.25 Billion Tonnes of CO_2 Under the North Seas," Dezeen, July 21, 2021, www.dezeen.com/2021/07/21/carbon-project-longship-norway-co2-north-sea/.

4. "vw-Chef Herbert Diess: 'Ich wünsche mir eine höhere $CO2$-Steuer von der Politik,'" *Wirtschaftswoche*, January 23, 2020, www.wiwo.de/unternehmen/industrie/autoindustrie-vw-chef-herbert-diess-ich-wuensche-mir-eine-hoehere-co2-steuer-von-der-politik/25467716.html.

5. David Ellison, Cindy E. Morris, Bruno Locatelli, et al., "Trees, Forests and Water: Cool Insights for a Hot World," *Global Environmental Change* 43 (March 2017): 51–61, doi.org/10.1016/j.gloenvcha.2017.01.002.

6. Daniel Eckert, "150.000.000.000.000 Dollar—der Wert des Waldes schlägt sogar den Aktienmarkt," *Die Welt*, August 8, 2020, www.welt.de/wirtschaft/article212771705/Neue-Studie-Waelder-der-Welt-sind-wervoller-als-der-Aktienmarkt.html.

19 | The Toilet Paper Argument

1. "BMELV: 1,3 Kubikmeter Holzverbrauch pro Kopf in Deutschland," press release from the Federal Ministry of Food and Agriculture, bonnsustainabilityportal.de/de/2012/09/bmelv-13-kubikmeter-holzverbrauch-pro-kopf-in-deutschland/.

2. Rheinland-Pfalz Ministerium für Umwelt, Energie, Ernährung, und Forsten, "Stickstoff im Wald—unverzichtbarer Nährstoff und waldgefährdender Schadstoff," Waldzustandesbericht 2016: 62–71,

mkuem.rlp.de/fileadmin/mulewf/Publikationen/Waldzustandsbericht_
2016.pdf.

3. Sophia Etzold, Marco Ferretti, Gert Jan Reinds, et al., "Nitrogen Depo-
sition Is the Most Important Environmental Driver of Growth of Pure,
Even-Aged and Managed European Forests," *Forest Ecology and Management* 458 (February 2020): 117762, doi.org/10.1016/j.foreco.2019.117762.

20 | More Money, Less Forest

1. Federal Ministry of Food and Agriculture, "Massive Schäden—Einsatz
   für die Wälder," www.bmel.de/DE/themen/wald/wald-in-deutschland/
   wald-trockenheit-klimawandel.html.

2. "Einschlagsmenge an Fichtenstammholz in Deutschland in den Jahren
   2000 bis 2020," Statista, September 2021, de.statista.com/statistik/
   daten/studie/162378/umfrage/ einschlagsmenge-an-fichtenstammholz-
   seit-1999/.

3. Amt für Ernährung, Landwirtschaft und Forsten Coburg-Kulmback,
   "Informationen zur Borkenkäfer und zur Bekämpfung," www.aelf-ck
   .bayern.de/forstwirtschaft/245181/index.php.

4. Fachagentur Nachwachsende Rohstoffe, "Ausgewählte Fördermaßna-
   hmen im Forstbereich von EU, Bund, Ländern und der Landwirtschaft-
   lichen Rentenbank," privatwald.fnr.de/foerderung#c39996.

5. Die Wald Eigentümer, "Unsere Landesverbände," www.waldeigen
   tuemer.de/verband/mitglieder/.

6. Josephine Andreoli, "Das verdienen die Abgeordneten aus dem
   Bundestag nebenher," abgeordnetenwatch.de, August 7, 2020,
   www.abgeordnetenwatch.de/blog/nebentaetigkeiten/das-verdienen-
   die-abgeordneten-aus-dem-bundestag-nebenher.

7. Die Wald Eigentümer, "Kooperation statt Verbote—Neustart beim
   Insektenschutz," www.waldeigentuemer.de/neustart-beim-
   insektenschutz/.

8. Fachagentur Nachwachsende Rohstoffe, "About," www.fnr.de/fnr-
   struktur-aufgaben-lage/fachagentur-nachwachsende-rohstoffe-fnr.

9. Fachagentur Nachwachsende Rohstoffe, "Der Brennstoff Holz,"
   heizen.fnr.de/heizen-mit-holz/der-brennstoff-holz/.

10. Fachagentur Nachwachsende Rohstoffe, "Mitglieder," www.fnr.de/
    fnr-struktur-aufgaben-lage/fachagentur-nachwachsende-rohstoffe-fnr/
    mitglieder.

11. In an article for the magazine *Öko-Test*, Maren Klein wrote: "Hinter dem
    PEFC-Label verbirgt sich ein Zertifizierungssystem von Forstindustrie

und Waldbesitzerorganisationen…Kaum eine Umweltorganisation unterstützt das PEFC-Label. Der WWF etwa hält das Waldzertifizierungssystem für 'nicht glaubwürdig.' " [Behind the PEFC label hides a certification system overseen by the timber industry and organizations of forest owners… Hardly any environmental organizations support the PEFC label. The WWF, for example, calls this timber certification system "not credible."] See: Maren Klein, "Waldsterben: Was jeder einzelne dagegen tun kann," Öko-Test, August 18, 2020, www.oekotest.de/freizeit-technik/Waldsterben-Was-jeder-einzelne-dagegen-tun-kann-_11401_1.html.

12. Bundesministerium für Ernährung und Landwirtschaft, Fachagentur Nachwachsende Rohstoffe, "Nachhaltigkeitsprämie Wald," www.bundeswaldpraemie.de/hintergrund.

13. Deutscher Bundestag, Drucksache 19/23755, October 28, 2020, dserver.bundestag.de/btd/19/237/1923755.pdf.

21 | The Ivory Tower Wobbles

1. Max-Planck-Gesellschaft, "Nachhaltige Wirtschaftswälder: ein Beitrag zum Klimaschutz," press release, February 10, 2020, www.mpg.de/14452850/nachhaltige-wirtschaftswalder-ein-beitrag-zum-klimaschutz.

2. This press release is no longer available on the Eberswalde University for Sustainable Development (HNEE) website, but you can find it here: Naturwald Akademie, "Waldschutz ist besser für das Klima als die Holz-Nutzung," naturwald-akademie.org/presse/pressemitteilungen/waldschutz-ist-besser-fuer-klima-als-holz-nutzung/.

3. Sebastiaan Luyssaert, Ernst-Detlef Schulze, Annette Börner, et al., "Old-Growth Forests as Global Carbon Sinks," Nature 455 (September 2008): 213–215, doi.org/10.1038/nature07276.

4. Department of Biogeochemical Processes at the Max Planck Institute for Biogeochemistry, "Carbon Balance, Ecosystem Research and Management, Prof. Dr. Dr.h.c Ernst Detlef Schulze," www.bgc-jena.mpg.de/bgp/index.php/EmeritusEDS/EmeritusEDS.

5. Keno Verseck, "Holzmafia in Rumänien—Förster in Gefahr," Spiegel, November 1, 2019, www.spiegel.de/panorama/justiz/holzmafia-in-rumaenien-zwei-morde-an-foerstern-a-1294047.html.

6. "Waldentwicklung im Nationalpark Hainich—Ergebnisse der ersten Wiederholung der Waldbiotopkartierung, Waldinventur und der Aufnahme der vegetationskundlichen Dauerbeobachtungsflächen," vol. 3

in the series *Erforschen der Tier-, Pflanzen- und Pilzwelt des Nationalparks* (Bad Langensalza: Nationalpark-Verwaltung Hainich, 2012).

7. Ernst-Detlef Schulze, Carlos A. Sierra, Vincent Egenolf, et al., "The Climate Change Mitigation Effect of Bioenergy From Sustainably Managed Forests in Central Europe," *Global Change Biology Bioenergy* 12, no. 3 (March 2020): 186–197, doi.org/10.1111/gcbb.12672.

8. The press release is no longer available on the website ofthe Eberswalde University for Sustainable Development (HNEE), but you can see it here: Naturwald Akademie, "Waldschutz ist besser für das Klima als die Holz-Nutzung," naturwald-akademie.org/presse/pressemit teilungen/waldschutz-ist-besser-fuer-klima-als-holz-nutzung/.

9. Thünen Institute, "Our Mission," www.thuenen.de/en/wf/mission-and-vision/.

10. Here's a link that includes the tweet from September 8, 2020. In 2021, the account was labeled Privat/Private and the author indicated he was using it to express his personal opinions: twitter.com/BolteAnd.

11. Bundesministerium für Ernährung und Landwirtschaft, "Wissenschaftlicher Beirat für Waldpolitik," www.bmel.de/DE/ministerium/ organisation/beiraete/waldpolitik-organisation.html.

12. The press release originally on the website of the Eberswalde University for Sustainable Development (HNEE) has been removed. The information can now be found here: Naturwald Akademie, "Waldschutz ist besser für das Klima als die Holz-Nutzung," naturwald-akademie.org/presse/pressemitteilungen/waldschutz-ist-besser-fuer-klima-als-holz-nutzung/.

13. Carpathia European Wilderness Preserve, "About," www.carpathia.org/ about/.

14. Pradip Krishen, "Introduction," *The Hidden Life of Trees* (Gurgaon: Penguin Random House India, 2016).

15. Marco Evers, "Wie ein Ölkonzern sein Wissen über den Klimawandel geheim hielt," April 16, 2018, *Spiegel*, www.spiegel.de/spiegel/wie-shell-sein-wissen-ueber-denklimawandel-geheim-hielt-a-1202889.html.

16. "Letter From Scientists to the EU Parliament Regarding Forest Biomass," updated January 14, 2018, www.pfpi.net/wp-content/ uploads/2018/04/UPDATE-800-signatures_Scientist-Letter-on-EU-Forest-Biomass.pdf.

17. Meghan O'Brien and Stefan Bringezu, "What Is a Sustainable Level of Timber Consumption in the EU: Toward Global and EU Benchmarks

for Sustainable Forest Use," *Sustainability* 9, no. 5 (2017): 812, doi.org/10.3390/su9050812.

18. "Erdölverbrauch in Europa in den Jahren von 1970 bis 2020," Statista, July 2021, de.statista.com/statistik/daten/studie/36202/umfrage/verbrauchvon-erdoel-in-europa/.

19. Bundesverfassungsgericht, Urteil vom 31.05.1990, NVwZ 1991, S 53.

20. Bundesverfassungsgericht, Beschluss des Zweiten Senats vom 12, Mai 2009–2 BvR 743/01–, Rn. 1–74, www.bundesverfassungsgericht.de/SharedDocs/Entscheidungen/DE/2009/05/rs20090512_2bvr074301.html.

21. Bundeskartellamt, "Holzverkauf ist keine hoheitliche Aufgabe," August 17, 2014, www.bundeskartellamt.de/SharedDocs/Interviews/DE/Stuttgarter_Ztg_Holzverkauf.html.

22. Marc Kubatta-Große. "Kartellklage gegen Forstministerium Rheinland-Pfalz," Forstpraxis, June 29, 2020, www.forstpraxis.de/kartellklage-gegen-forstministerium-rheinland-pfalz.

23. Marc Kubatta-Große, "Kartellklage gegen Forstministerium Rheinland-Pfalz," Forstpraxis, June 29, 2020, www.forstpraxis.de/kartellklage-gegen-forstministerium-rheinland-pfalz.

22 | What's on Your Plate?

1. Bundesinformationszentrums Landwirtschaft, "Was wächst auf Deutschlands Feldern?" www.landwirtschaft.de/landwirtschaft-verstehen/wie-arbeiten-foerster-und-pflanzenbauer/was-waechst-auf-deutschlands-feldern.

2. "Der Ökowald als Baustein einer Klimaschutzstrategie," Gutachten im Auftrag von Greenpeace e.V., www.greenpeace.de/sites/www.greenpeace.de/files/publications/20130527-klima-wald-studie.pdf.

3. "Kohlenstoffspeicherung von Bäumen," Bayerische Landanstalt für Wald und Forstwirtschaft, Merkblatt 27, July 2011, www.lwf.bayern.de/mam/cms04/service/dateien/mb-27-kohlenstoffspeicherung-2.pdf.

4. Bundesministerium für Ernährung und Landwirtschaft, "Holzzuwachs auf hohem Niveau," in *Dritte Bundeswaldinventur* 2012, www.bundeswaldinventur.de/dritte-bundeswaldinventur-2012/rohstoffquelle-wald-holzvorrat-auf-rekordniveau/holzzuwachs-auf-hohem-niveau/.

5. "Methan: Wie Rinder dem Klima schaden," Wirtschaftswoche, March 27, 2017, www.wiwo.de/technologie/green/methan-wie-rinder-dem-klimaschaden/19575014.html.

6. "Das steckt hinter einem Kilogram Rindfleisch," Albert Schweitzer Stiftung,November 1, 2016, albert-schweitzer-stiftung.de/aktuell/1-kg-rindfleisch.

7. Statistik des Bundesministeriums für Ernährung und Landwirtschaft für das Jahr 2019, www.bmel-statistik.de/ernaehrung-fischerei/versorgungsbilanzen/fleisch/.

8. Umwelt Bundesamt, "Treibhausgas-Ausstoß pro Kopf in Deutschland nach Konsumbereichen (2017)," www.umweltbundesamt.de/bild/treibhausgas-ausstoss-pro-kopf-in-deutschland-nach.

9. "Life-Cycle Analysis Study Suggests Eating Less Meat," Food Engineering, July 9, 2012, www.foodengineeringmag.com/articles/89503-life-cycle-analysis-study-suggests-eating-less-meat.

10. Norbert Lehmann, "Niederlande bieten Ausstiegsprämie für Tierhalter an," Agrarheute, November 4, 2020, www.agrarheute.com/politik/niederlande-bieten-ausstiegspraemie-fuer-tierhalter-574652.

11. Statistik des Bundesministeriums für Ernährung und Landwirtschaft für das Jahr 2019, www.bmel-statistik.de/ernaehrung-fischerei/versorgungsbilanzen/fleisch/.

12. Gesetz über den Nationalpark Unteres Odertal, Gesetz und Verordnungsblatt für das Land Brandenburg, Potsdam, November 16, 2006.

13. "Es gibt wieder frei lebende Wisente," Wisent Welt, www.wisent-welt.de/artenschutz-projekt.

14. Ulrike Fokken, "Wildtiere im Rothaargebirge: Ein 900 Kilo schweres Problem," Taz.de, May 24, 2020, taz.de/Wildtiere-im-Rothaargebirge/!5684424/.

23 | Every Tree Counts

1. François-Alain Daudet, Thierry Améglio, Hervé Cochard, et al., "Experimental Analysis of the Role of Water and Carbon in Tree Stem Diameter Variations," Journal of Experimental Botany 56, no. 409 (January 2005): 135–144, doi.org/10.1093/jxb/eri026.

2. Marion Zapater, Christian Hossann, Nathalie Bréda, et al., "Evidence of Hydraulic Lift in a Young Beech and Oak Mixed Forest Using [18]O Soil Water Labelling," Trees 25, no. 5 (2011): 885–894, doi.org/10.1007/s00468-011-0563-9.

3. Todd E. Dawson, "Hydraulic Lift and Water Use by Plants: Implications for Water Balance, Performance and Plant-Plant Interactions," Oecologia 95 (1993): 565–574, doi.org/10.1007/BF00317442.

24 | Does Everyone Have to Be On Board?

1. Georg Sperber and Norbert Panek, "Was Aldo Leopold sagen würde," in *Der Holzweg* (Munich: Oekom Verlag, 2021): 81–90, cdn.website-editor.net/93acd78a5ec24fa1ba52fe5e3c85ea1d/files/uploaded/Der%2520Holzweg.pdf.

2. Wolfgang, "GRÜNE LIGA Sachsen und NUKLA ./. Stadt Leipzig: Beschluss des OVG Bautzen vom 09.06.2020," Naturschutz und Kunst—Lebendige Auen e.V., www.nukla.de/202%/gruene-liga-sachsen-und-nukla-stadt-leipzig-beschluss-des-ovg-bautzen-vom-9-6-2020/.

3. Georg Becher, "Clusterstatistik Forst und Holz: Tabellen für das Bundesgebiet und die Länder 2000 bis 2013," Thünen Working Paper 48 (October 2015): 14, www.thuenen.de/media/publikationen/thuenen-workingpaper/ThuenenWorkingPaper_48.pdf.

4. *The Hidden Life of Trees*, directed by Jörg Adolph and Jan Haft (Constantin Film, 2020).

25 | A Breath of Fresh Air

1. Hochschule für Forstwirtschaft Rottenburg, "Hochschulen (HAW / FH) und Universitäten mit forstlichen Studienangeboten in Deutschland," March 9, 2021, www.hs-rottenburg.net/aktuelles/aktuelle-meldungen/meldungen/aktuell/2021/gemeinsame-erklaerung/.

26 | The Forest Will Return

1. Tina Baier and Marlene Weiss, "Es ist nicht der Wald, der stirbt, es sind die Bäume," *Stuttgarter Zeitung* 228 (October 2, 2020): 36–37.

2. Bündesministerium für Umwelt, Naturschutz, nukleare Sicherheit und Verbraucherschutz, "FAQ: Was ist das Schutzgebietnetz Natura 2000 in Deutschland?" www.bmuv.de/faq/was-ist-das-schutzgebietsnetz-natura-2000-in-deutschland.

3. "Warum Wildnis," Wildnis in Deutschland, wildnisindeutschland.de/warum-wildnis/.

4. DePaul University, "Symbiotic Underground Fungi Disperse by Wind, New Study Finds," press release, July 7, 2020, resources.depaul.edu/newsroom/news/press-releases/Pages/dispersal-of-mycorrhizal-fungi-.aspx.

5. Olaf Spörkel, "Überraschend hohe Anzahl an Pilzsporen in der Luft," Laborpraxis, July 14, 2009, www.laborpraxis.vogel.de/ueberraschend-hoheanzahl-an-pilzsporen-in-der-luft-a-200852.

Index

DAVID SUZUKI INSTITUTE

THE DAVID SUZUKI INSTITUTE is a nonprofit organization founded in 2010 to stimulate debate and action on environmental issues. The Institute and the David Suzuki Foundation both work to advance awareness of environmental issues important to all Canadians.

We invite you to support the activities of the Institute. For more information please contact us at:

David Suzuki Institute
219 – 2211 West 4th Avenue
Vancouver, BC, Canada V6K 4S2
info@davidsuzukiinstitute.org
604-742-2899
davidsuzukiinstitute.org

Checks can be made payable to The David Suzuki Institute.